A Cascade of Numbers

A Cascade of Numbers

An Introduction to

Number Theory

arranged by

Bob Burn
School of Education
University of Exeter

Amanda Chetwynd
Department of Mathematics and Statistics
Lancaster University

A member of the Hodder Headline Group
LONDON • SYDNEY • AUCKLAND

First published in Great Britain 1996 by Arnold,
a member of the Hodder Headline Group,
338 Euston Road, London NW1 3BH

British Library Cataloguing in Publication Data
A catalogue record for this book is available from the British Library

ISBN 0 340 65251 9

Typeset in 10/12pt Galliard by Graphicraft Typesetters Ltd, Hong Kong
Printed and bound in Great Britain by J W Arrowsmith Ltd, Bristol

Contents

		Comments and Solutions
Preface	vii	
Definitions	ix	
Part I	1	75
1. The magic of nines (decimal place value)	3	77
2. Back to basics (bicimal place value)	5	78
3. Children's ages (prime factorisation part i)	5	79
4. The prison door problem (counting factors)	5	79
5. Catching practice (repeated addition around a circle part i)	7	81
6. Algorithma (highest common factor, part i)	8	81
7. Breeding rabbits (Fibonacci, using factors)	10	82
8. To divide or not to divide (prime factorisation part ii)	12	83
9. A return visit to Algorithma (highest common factor, part ii, the fundamental theorem of arithmetic)	13	84
10. The stamp problem	13	85
11. Tests for divisibility (decimal place value)	15	86
12. Spot check (first taste of *modulo* 10)	15	
13. Eratosthenes' sieve (locating primes, proof by contradiction)	17	87
14. Raffle tickets and neighbours (primes and composites)	17	88
15. How many primes? (no finite number, contradiction again)	18	88
16. No shuffling (first taste of *modulo* 4)	19	89
17. Can you really tell the time? (*modulo* 12 and *modulo* 4)	19	90
18. Fibonacci numbers and the division algorithm	21	92
19. Relations can be difficult (equivalence)	22	93
20. Dominoes (induction)	25	95
Part II	29	99
21. Chinese remainders	31	101
22. Systematic catching practice (repeated addition part ii, ϕ)	32	102
23. Do you know your tables? (modular multiplication)	33	104
24. Coding and decoding (simple methods)	34	107
25. Repacking (squares and their residues)	35	107
26. Where have all the squares gone?	36	108
27. Where have all the squares come from?	36	109
28. How old is Grandma? (applying quadratic residues)	37	110
29. Higher powers (approaching Fermat's theorem)	39	111
30. Just shuffling and then ^{power} (*modulo* 7) (Fermat's theorem part i)	40	112
31. Just shuffling and then ^{power} (*modulo* 3, 5, 11) (Fermat's theorem part ii)	41	113

	Comments and Solutions

32. Factorials (Wilson's theorem)   43   114
33. Square roots of -1, prime modulus ($p \equiv 1, 3 \pmod 4$)   43   115
34. How many square roots of -1? (factors of $x^2 + 1$)   44   115
35. Sums of squares   44   116
36. Sums of squares in two ways   46   117
37. Pythagorean triples   47   118

Part III   49   121
38. Squares and non-squares (quadratic residues and products)   51   123
39. Powers of squares and non-squares (towards the Legendre
    symbol)   52   123
40. The frequency of factors ($\Sigma\phi(d)$)   52   124
41. Multiplication like addition (cyclic groups and generators)   53   125
42. Powers to a prime modulus (primitive roots)   54   126
43. Zero products (*modulo* a non-prime)   56   127
44. Non-zero products (Fermat–Euler theorem)   57   128
45. Decimals to the death (recurring decimals to fractions)   57   129
46. Recurring decimals (fractions to recurring decimals)   58   131
47. Can you reveal all and keep it secret? (Public Key System)   60   133
48. Primes as squares and non-squares (starting quadratic
    reciprocity)   62   134
49. Counting dots in a rectangle (odd or even)   64   135
50. Half-size products (Gauss' lemma, when 2 is a square)   64   136
51. From dashes to dots (Eisenstein)   66   138
52. Quadratic reciprocity   68   141
53. Adding squares (primes in various forms)   70   143

Index   147

Preface

What are the facts?

▷ Students of mathematics start their higher education with a wide variety of backgrounds; and

▷ there are persistent calls to widen access to higher education.

British universities need to respond by

▷ offering courses in mathematics which are comprehensible from the start whatever A-level or access route students may have taken;

▷ designing mathematics courses which consciously lead students towards more rigorous proofs but do not assume that students have an innate love of rigour;

▷ helping students to work with and learn with each other;

▷ offering a more participatory learning style than block lectures.

This book is designed to satisfy these needs for a course in number theory. It is divided into 53 sections and each section has two halves. The first half of each section is raw material encouraging student activity that leads to pattern recognition and conjecture. In the second half of each section, along with comments and solutions, formal structures are identified and proofs are constructed.

A variety of number theoretic strands are explored in the sections:

Place value in sections 1, 2, 11, 12.
Prime numbers in sections 3, 4, 8, 9, 13, 15.
Factors in sections 4, 5, 6, 7, 8, 14, 15, 18, 43, 45.
Congruence in sections 5, 6, 13, 16, 17, 19, 21, 23.
Euler's ϕ function in sections 5, 22, 24, 40.
Quadratic residues in sections 25, 26, 27, 28, 33, 34, 35, 36, 37, 38, 39, 48, 49, 50, 51, 52, 53.
Fermat and Fermat–Euler in sections 29, 30, 31, 42, 44, 46, 47.

Methods of proof are initiated at different stages:

Generality through algebra in section 11.
Exhaustion in section 12.
Contradiction in section 13.
Induction in sections 9 and 20.

There is intended overlap, especially in the first 20 sections, but enough variety in these materials either for a post-GCSE or a post-A-level course. The variety here will support either stand-alone modules for arts students, courses for teachers in training or a first course in number theory for those specialising in mathematics. Part I (sections 1–20) could be part of a sixth-form course and we expect the lecturer to make a selection from this material. Part II (sections 21–37) is intended for first year undergraduate study and Part III (sections 38–53) for second year courses at university. In Parts II and III the main development (sections 21–23, 26, 27, 29–33,

35, 38–44, 48–53) leads towards quadratic reciprocity and there are excursions into coding, factorisation and decimals (sections 24, 25, 28, 36, 45–47). As the section numbers rise, the amount of overlap diminishes and the dependence on earlier sections grows.

Teaching style 1, for smaller groups

The raw material in the first half of the sections is used for group work in class. The lecturer can intervene from time to time with comments from the second half of the section. Homework consists of individual investigations, running alongside the classwork provided by the text. The combination of group work and individual investigations provides the basis for identifying mathematical processes: an initial question or situation, the gathering of data, the organising and display of data, the recognition of pattern, the description of pattern, the formulation of conjectures, prediction and the testing of conjectures, the construction of counter-examples, the revision of conjectures, justification and proof, the usefulness of definitions.

Teaching style 2, for larger groups

The material fits the conventional lecture/exercise format, provided the usual sequence is reversed, and the exercises in the first half of each section are set to be completed **before** the lecture which is to deal with the related comment in the latter half of the section. The students will then have seen the ideas at work before the theory is developed by lecture and discussion.

Definitions

1, 2, 3, 4, . . .   **N**, the natural numbers, counting numbers, positive integers.

0, ±1, ±2, ±3, . . .   **Z**, the integers.

When a, b and c are integers, and $a = b \cdot c$, we say

> b is a factor of a, or
> b divides a, or
> b is a divisor of a, or
> a is a multiple of b,

each of which is expressed symbolically by writing $b \mid a$.

For positive integers, if $d \mid a$ and $d \mid b$, d is called a common factor or common divisor of a and b. If h is a common factor of a and b and every common factor of a and b divides h, then h is called the *highest common factor* or *greatest common divisor* of a and b, and we write $h = \text{hcf}(a, b)$ or $\gcd(a, b)$. When the highest common factor of two positive integers is 1, the integers are said to be *coprime*. When $\text{hcf}(a, b) = 1$, it is commonly, though not correctly, said that "a and b have no common factor".

The integer p is said to be a *prime number*, when p is positive, not equal to 1, and the only positive divisors of p are 1 and p itself.

$\phi(n)$ = the number of integers in the set $\{a \mid 1 \le a \le n \text{ and } \text{hcf}(a, n) = 1\}$.

When $n \mid a - b$, we write $a \equiv b \pmod{n}$ and say that a is congruent to b *modulo n*.

If for some integer x, $x^2 \equiv a \pmod{n}$, then a is called a square or quadratic residue modulo n. If there is no integer x such that $x^2 \equiv a \pmod{n}$ then a is called a non-square or quadratic non-residue modulo n.

For any real number x, $\lfloor x \rfloor$ (said, floor x) is the integer such that $\lfloor x \rfloor \le x < \lfloor x \rfloor + 1$. For example, $\lfloor 22/7 \rfloor = 3$.

PART I

1 The magic of nines

In the nine times table below notice that the digits of each product sum to 9. Why does this happen? Look at how the digits of the product are changing each time.

$$1 \times 9 = \ 9$$
$$2 \times 9 = 18$$
$$3 \times 9 = 27$$
$$4 \times 9 = 36$$
$$5 \times 9 = 45$$
$$6 \times 9 = 54$$
$$7 \times 9 = 63$$
$$8 \times 9 = 72$$
$$9 \times 9 = 81$$
$$10 \times 9 = 90$$

Think of a two digit number, say 42, then subtract the reverse of its digits, 24, from 42. Choose three more two digit numbers and for each one reverse the digits and subtract the smaller number from the larger. Look at all the answers you get. Do they all have a common divisor? What do the digits sum to each time?

Can you show that the answer will always have a factor of 9? Think of the first two digit number as ab where $a > b$, then reverse it to get ba and subtract ba from ab. Show that the answer has a factor of 9. Can you also show that the digits will always sum to 9?

Think of a three digit number, say 524, then subtract its reversal 425 from 524. Choose three more three digit numbers and for each one reverse it and subtract the smaller number from the larger. Look at all the answers you get and again see if you can find a common divisor and also see what the sum of the digits is each time.

If the first three digit number is abc where $a > c$, why does $abc - cba$ have a factor of 99 and its digits sum to 18? Now suppose $abc - cba = pqr$, then what is $pqr + rqp$?

2332, 5445, 6776 and 4224 are numbers which read the same forwards and backwards. Such numbers are called *palindromic* numbers with four digits. Find a common factor of them all. Write down another palindromic number with four digits and check that it too is divisible by the same number. Will this always be the case?

2 Back to basics

There is a new head of mathematics at Trumpton School. She has upset all the old teachers by saying that the pupils do not need to learn their multiplication tables and only have to be able to multiply and divide by 2. The teachers ask: then how can the pupils be expected to multiply numbers like 17 and 31? The head was ready to accept the challenge. She drew up a table with two columns labelled halving and doubling. At the top of the first column she wrote 17 and underneath each time wrote half the number above (ignoring the remainder if there is one). So underneath 17 she wrote 8 and continued halving until the number 1 was reached. In the column to the right of the 17 she wrote the other number to be multiplied, 31, and then underneath wrote its double and continued doubling as many times as she had halved.

"Halving"	"Doubling"
17	31
8	62
4	124
2	248
1	496

Now to perform the multiplication, she crossed out any number in the doubling column that appeared beside an *even* number in the halving column. Then she added the numbers which remained in the doubling column. The result, she announced, will be equal to the product of the two numbers she started with.

Was she right?

Try this for another pair of numbers, 57 and 23. Check your answer by normal multiplication. Can you see why it works?

For each of the figures still left in the doubling column notice how many times 31 has been doubled.

Make 17 by selecting and adding single numbers from the list 1, 2, 3, 4, 8, 16, $32, 64, \ldots, 2^n, \ldots$:

$$17 = 1 \cdot 16 + 0 \cdot 8 + 0 \cdot 4 + 0 \cdot 2 + 1 \cdot 1$$

Try to see how the repeated halving of 17 gives the zeros and ones in this expression.

We shall now see how expressing numbers as powers of 2 is used as a basis for a magic trick.

A mind reading trick

A set of mind reading cards:

1	3	5	7
9	11	13	15
17	19	21	23
25	27	29	31
33	35	37	39
41	43	45	47
49	51	53	55
57	59	61	63

2	3	6	7
10	11	14	15
18	19	22	23
26	27	30	31
34	35	38	39
42	43	46	47
50	51	54	55
58	59	62	63

4	5	6	7
12	13	14	15
20	21	22	23
28	29	30	31
36	37	38	39
44	45	46	47
52	53	54	55
60	61	62	63

8	9	10	11
12	13	14	15
24	25	26	27
28	29	30	31
40	41	42	43
44	45	46	47
56	57	58	59
60	61	62	63

16	17	18	19
20	21	22	23
24	25	26	27
28	29	30	31
48	49	50	51
52	53	54	55
56	57	58	59
60	61	62	63

32	33	34	35
36	37	38	39
40	41	42	43
44	45	46	47
48	49	50	51
52	53	54	55
56	57	58	59
60	61	62	63

A magician asks you to choose a number between 1 and 63 and point to the cards on which your number occurs. You choose a number (say 27) and point to the cards with 27 on. The magician notices that the cards pointed to have 1 2 8 16 in the top left hand corner. He announces your number was 27!

How does the magician do it?

The magician just has to add the numbers in the top left corner of all the cards that the number is on and this gives the number. Try it with another number.

Can you determine how the cards are constructed? What numbers are in the top left hand corners? Why were these particular numbers chosen?

3   Children's ages

A friend tells me her children's ages are all primes and multiply together to give 595. That's strange because my children's ages are also primes and when I work out the product of their ages I also get 595. My friend says that we must have children of the same age. Is she right and if so what are their ages?

Another friend remarks that his children's ages multiply to 48. Can you say what age his children are?

4   The prison door problem

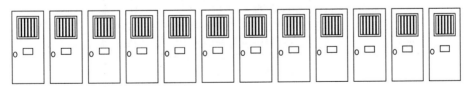

Figure 1 Prison doors

The doors of 10 prison cells are all closed. Ten warders come along, one by one.

Warder No. 1 opens all 10 doors.

Warder No. 2 then arrives and closes every other door, starting with door number 2.

Warder No. 3, starting with door number 3 and dealing only with every third door, opens those which are shut and closes those which are open.
Warder No. 4, starting with door number 4 and dealing only with every fourth door, opens those which are shut and closes those which are open.

The process continues with the arrival of each new warder up to Warder No. 10. Which prison doors are eventually left open and which are finally shut?

Now try the same problem but with 100 prison cells which all start closed and 100 warders come along, one by one, opening and shutting doors as before.

Warder No. 1 opens all hundred doors.
Warder No. 2 then arrives and closes every other door, starting with door number 2.
Warder No. 3, starting with door number 3 and dealing only with every third door, opens those which are shut and closes those which are open.
Warder No. 4, starting with door number 4 and dealing only with every fourth door, opens those which are shut and closes those which are open.

The process continues with the arrival of each new warder up to Warder No. 100. Which prison doors are eventually left open and which are finally shut?

The state of prison door number 15 is changed by Warders Nos 1, 3, 5 and 15, and no more, so prison door number 15 is finally shut. The warders who open or close a door are those whose number is a factor of the door number. If we count how many factors of a door number there are, we find out how many warders open or close that door. Having found the number of factors we can determine whether the number of factors is odd or even. An even number of factors results in a closed door. An odd number of factors results in an open door.

If a door number has exactly two factors, the door is finally shut. Find all the numbers up to 30 which have exactly two factors. What do we call these numbers? See the comment on **section 3**.

If a door number has exactly three factors, the door is eventually left open. Find all the numbers up to 100 which have exactly three factors. In what way are these numbers like each other?

The number 12 has six factors:
1, 2, 3, 4, 6 and 12.
These are exhibited in the lattice.
A vertical bar separates factors
differing by a factor of 2, a
diagonal bar separates factors
differing by a factor of 3.

Draw analogous lattices for 15 and for 18, using a different direction for the two prime factors of each number.

List as many numbers as you can, under a hundred, which have each of the following factor lattices:

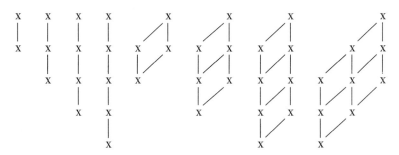

Although $245 = 5 \cdot 7^2$ is greater than 100, you may be able to tell which of the above factor lattices it has, and therefore be able to deduce how many factors it has.

Express each number in the lattice for 12 in the form $2^a \cdot 3^b$.

If $5^a \cdot 7^b$ is a factor of $5^6 \cdot 7^3$ what possible values may a and b take? So how many factors does $5^6 \cdot 7^3$ have? Is this an odd or even number?

How many factors does 2^x have?
How many factors does $2^x \cdot 3^y$ have?
How many factors does $2^x \cdot 3^y \cdot 5^z$ have?
How many factors does $2^x \cdot 3^y \cdot 5^z \cdot 7^t$ have?

How do you look for a number with an odd number of factors?

5 Catching practice

Twelve cricketers are standing in a circle. One is holding a ball. The one with the ball throws it to the person three away to his left (i.e. passing two people). The catcher does the same, and so on. Does every cricketer touch the ball? Draw a circle; mark the positions of the players around the circle and, with a pencil, show the path of the ball as it is thrown from person to person.

If the players throw to the person four away to the left (i.e. passing three players) does every cricketer touch the ball this time?

Try steps of other lengths.

You can also think of what is happening in terms of the hour hand on a clock face. If it were repeatedly moved on x hours would the hand point to all the hours, or not?

Make a list of the step lengths by which every cricketer gets a touch and a list of the step lengths which do not give every cricketer a touch of the ball. What is the difference between the numbers in these two lists? Can you describe a property which holds together all the numbers in one list and distinguishes them from all the numbers in the other?

Now try with nine cricketers standing in a circle. What are the step lengths which reach every cricketer? What are the step lengths which do not? Does the property which you chose to distinguish the two lists in the case of 12 cricketers adequately describe the two lists which you get in the case of nine cricketers?

Apply the property you have chosen to describe the difference between the step lengths which reach every person around a circle and the step lengths which do not reach every person, to predict what step lengths would reach every person if there were 11 cricketers standing in a circle and the ball was thrown round using a constant gap.

6 Algorithma

A long time ago, before the days of coins and bank notes, the inhabitants of the beautiful tropical island of Algorithma had well established conventions of fairness when goods were exchanged. Sugar cane was plentiful on the island and there were quite a few coconut palms. Some of the inhabitants were expert ocean-fishermen. A coconut was regularly exchanged for five sticks of sugar cane, and a large fish was exchanged for 12 sticks of sugar cane.

I had been fishing all day and had returned to the island with my catamaran heavily laden with fish. I was desperate to chew some sugar cane. "Please just give me one stick", I said to the first friend I met whom I knew to have a good store of both sugar cane and coconuts. "Fair's fair" he said, "you give me some fish, and I'll give you coconuts and sugar cane." "But I only want one stick of sugar cane", I said. "OK" he said, "then we'll have to work out the transaction."

He was seeking a fair exchange in giving just one stick of sugar cane. How many coconuts must be put with a stick of sugar cane to make a fair exchange for some whole fish?

There are several possible answers. Find two or three before reading further.

Circle the number of sticks of sugar cane which may be exchanged for a whole number of fish on the table displayed below:

1	2	3	4	5
6	7	8	9	10
11	12	13	14	15
16	17	18	19	20
21	22	23	24	25
26	27	28	29	30
31	32	33	34	35
36	37	38	39	40
41	42	43	44	45
46	47	48	49	50
51	52	53	54	55
56	57	58	59	60

Using the display above and starting from a single stick of sugar cane, what are the numbers of sticks of sugar cane which are worth one stick of sugar cane with one coconut, worth one stick of sugar cane with two coconuts, and so on? Do you eventually reach a number which can be exchanged for a whole number of fish?

Another time, I had spent the whole day climbing coconut palms for their fruit and, after the day's work, was desperate for sugar cane to chew. This time my friend with the store of sugar cane had been fishing, and a similar conversation took place. What transaction could satisfy my needs and give me one stick of sugar cane? On the array of numbers of sugar cane below, circle the numbers of sticks of sugar cane which may be exchanged for a whole number of coconuts.

1	2	3	4	5	6	7	8	9	10	11	12
13	14	15	16	17	18	19	20	21	22	23	24
25	26	27	28	29	30	31	32	33	34	35	36
37	38	39	40	41	42	43	44	45	46	47	48
49	50	51	52	53	54	55	56	57	58	59	60

Starting with a single stick of sugar cane, what are the numbers of sticks of sugar cane which are worth one stick of sugar cane with one fish, worth one stick of sugar cane with two fish, and so on? Can you reach a number which can be exchanged for a whole number of coconuts?

One way of doing a systematic search might go like this:

$$12 \cdot s = 1 \cdot F$$
$$5 \cdot s = 1 \cdot C$$
$$12 \cdot s - 5 \cdot s = 7 \cdot s = 1 \cdot F - 1 \cdot C$$
$$7 \cdot s - 5 \cdot s = 2 \cdot s = (1 \cdot F - 1 \cdot C) - 1 \cdot C = 1 \cdot F - 2 \cdot C$$
$$5 \cdot s - 2 \cdot s = 3 \cdot s = 1 \cdot C - (1 \cdot F - 2 \cdot C) = -1 \cdot F + 3 \cdot C$$
$$3 \cdot s - 2 \cdot s = 1 \cdot s = (-1 \cdot F + 3 \cdot C) - (1 \cdot F - 2 \cdot C) = -2 \cdot F + 5 \cdot C$$

So I may exchange five coconuts for two fish and one stick of sugar cane.

Another way to see what is happening here is to work with pairs of numbers, starting with 12 and 5. If we keep subtracting the lesser from the greater we get

$$[12, 5] \rightarrow [7, 5] \rightarrow [2, 5] = [5, 2] \rightarrow [3, 2] \rightarrow [1, 2] = [2, 1] \rightarrow [1, 1]$$

When the two numbers are the same we stop, and by working backwards we can find the last number as a combination of the two we started with:

$$\begin{aligned}
1 &= 3 - 2 \\
&= (5 - 2) - 2 \\
&= 5 - 2 \cdot 2 \\
&= 5 - 2 \cdot (7 - 5) \\
&= 3 \cdot 5 - 2 \cdot 7 \\
&= 3 \cdot 5 - 2 \cdot (12 - 5) \\
&= 5 \cdot 5 - 2 \cdot 12
\end{aligned}$$

These equations are constructed using the first appearance of each number in the chain.

Devise some other rules of exchange (i.e. choose numbers other than 12 and 5 for fish and coconuts) which would be as discriminating (i.e. which could be combined in some way equivalent to one stick of sugar cane) as the one we used. What about 7 and 10, or 4 and 6?

If a coconut was normally exchanged for 15 cowrie shells and a fish for 33 cowrie shells, is there any way I could balance a fish–coconut transaction with a single cowrie shell?

Try working with the pair [33, 15] and repeatedly subtracting the smaller from the larger. What is the number when you have reached the point where both numbers in the pair are the same?

If you did the same exercise starting with a pair like [63, 49] what number do you think you would get when you have reached the point where the two numbers are the same? Does any special name come to mind?

If you start with the number pair [a, b] and then subtract the smaller number from the larger number to get [$a - b$, b], say, can you be sure that any factor of a and b has to be a factor of $a - b$ and b? Conversely, can you be sure that any factor of $a - b$ and b is a factor of a and b? So now, if you start with [a, b] and finally reach [c, c], how might you describe c in relation to a and b?

Must there be integers x and y such that $c = xa + yb$?

Must $xa + yb$ be divisible by any common factor of a and b?

7 Breeding rabbits

Figure 2 A rabbit in a hat

Leonardo of Pisa, commonly known as Fibonacci (son of Bonacci), a twelfth/thirteenth century Italian mathematician, posed the following problem in his book, *Liber Abaci*, of 1202.

One pair of rabbits is put in a certain place surrounded by a wall. How many rabbits can be produced from that pair in a year, if the nature of these rabbits is such that every month each pair bears a new pair which from the second month on becomes productive?

Complete the table below.

Month	Adult pairs	New-born pairs	Total pairs
1	1	1	2
2	2	1	3
3	3	2	5
4			
5			
6			

Consider the sequence

$$1, 1, 2, 3, 5, 8, \ldots$$

Can you give the next term in the sequence, and the next, and the next . . . ?

This sequence is called the Fibonacci sequence. It has many interesting properties and we shall explore some of them. It is still of interest today and there is a research journal called *The Fibonacci Quarterly* which has articles about new properties or applications of this sequence.

Can you see how the Fibonacci sequence relates to the rabbit problem? If we describe the terms of the sequence as $f(1) = 1, f(2) = 1, f(3) = 2$, and so on, we can see how $f(3)$ depends on $f(1)$ and $f(2)$:

$$f(1) + f(2) = \ldots ,$$

and how $f(4)$ depends on $f(2)$ and $f(3)$:

$$f(2) + f(3) = \ldots$$

In general the Fibonacci sequence is given by the following rules:

$$f(1) = 1, f(2) = 1, f(n + 2) = f(n + 1) + f(n) \text{ for } n = 1, 2, 3, \ldots$$

The first 30 Fibonacci numbers

1, 1, 2, 3, 5, 8, 13, 21, 34, 55, 89, 144, 233, 377, 610, 987, 1597, 2584, 4181, 6765, 10946, 17711, 28657, 46368, 75025, 121393, 196418, 317811, 514229, 832040.

Highest common factors

Look at the list of the first 30 Fibonacci numbers. Choose a pair of consecutive terms and see what their highest common factor is. Choose another pair of numbers and find their highest common factor. Would you expect this to be the same for any consecutive pair? Can you prove it?

Again using the list take several pairs of Fibonacci numbers and work out the highest common factor for each pair. Put your results in the following table. From your results can you see any relation between the Fibonacci numbers and their highest common factors?

Fibonacci number	Fibonacci number	Highest common factor
$610 = f(15)$	$144 = f(12)$	$2 = f(3)$

What Fibonacci number do you expect to be the highest common factor of $f(600)$ and $f(56)$?

8 To divide or not to divide

In the following table, 420 has been written as a product of two numbers in every possible way across the top and the proper divisors of 420 are written down the side.

	21·20	28·15	30·14	35·12	42·10	60·7	70·6	84·5	105·4	140·3	210·2
2	yes	yes	yes	yes							
3	yes	yes	yes	yes							
4	yes	yes	no	yes							
5	yes	yes	yes	yes							
6	no	no	yes	yes							
7											
10											
12											
14											
15											
20											
21											
28											
30											
35											
42											
60											
70											
84											
105											
140											
210											

For each divisor and product enter *yes* if the divisor on the left divides one of the terms of the product and *no* otherwise. Complete the entries in the table. Are there any rows with *yes* occurring right across? What is special about the left hand number in these cases? Why do some numbers always divide one of the terms no matter which way you write the product?

9   A return visit to Algorithma

On Algorithma (**section 6**) one fish was exchanged for 12 sticks of sugar cane, and one coconut was exchanged for five sticks of sugar cane. We worked with a transaction which used a single stick of sugar cane, noticing that five coconuts may be exchanged for two fish and one stick of sugar cane. The fairness of this transaction was based on the equation

$$5 \cdot 5 - 2 \cdot 12 = 1$$

Find some other transactions of coconuts and fish which are fair if one stick of sugar cane is also exchanged at the same time. Try to find a family of such transactions, and then look for a rule for generating them.

> We are really looking for integers x and y such that $x \cdot 5 + y \cdot 12 = 1$.
> Since we already know that $\qquad\qquad 5 \cdot 5 + (-2) \cdot 12 = 1$,
> it follows that $\qquad\qquad\qquad (x - 5) \cdot 5 + (y + 2) \cdot 12 = 0 \qquad\qquad (*)$

Now use the idea of **section 8** to say why $y + 2$ must be a multiple of 5, so that $y + 2 = 5 \cdot t$ for some integer t.

Substitute $y = 5 \cdot t - 2$ in equation $(*)$ to find x in terms of t. Do these values of x and y satisfy the equation $x \cdot 5 + y \cdot 12 = 1$, whatever the value of t?

Find some transactions of coconuts and fish which are fair if two sticks of sugar cane are exchanged at the same time. Devise a transaction of coconuts and fish which is fair if k sticks of sugar cane are exchanged at the same time.

On Algorithma, one fish was exchanged for 33 cowrie shells, and one coconut was exchanged for 15 cowrie shells. Find a transaction of coconuts and fish which is fair if three cowrie shells are also exchanged at the same time. Find a family of such transactions.

Is it possible to find a transaction of coconuts and fish which is fair if one cowrie shell is also exchanged at the same time? What about two cowrie shells? What numbers of cowrie shells can possibly balance a coconut–fish transaction?

10   The stamp problem

Once upon a time, second class stamps cost 5p and first class stamps cost 7p. If you had an unlimited supply of stamps of these two denominations what postage values could you stick on a packet? 12p can be done because $12 = 7 + 5$. 11p cannot. See what you can find out about this situation before reading further.

If you just focus on what you can do with 5p stamps, there is a simple but useful way of displaying the possibilities:

1	2	3	4	5
6	7	8	9	10
11	12	13	14	15
16	17	18	19	20
21	22	23	24	25
26	27	28	29	30
31	32	33	34	35
36	37	38	39	40
. . .	. . .	. . .	. . .	. . .

Every value in the last column can be reached using 5p stamps. No value in any other column can be reached using 5p stamps alone. This display is particularly useful for the further analysis of this problem, not just because every positive whole number has a unique place in this display (if it is carried on downwards without limit), but also because the step from a number to the number directly below it is taken by adding 5. This means that once a number has been reached using 5p and 7p stamps, every number below it can be reached by adding on 5p stamps.

Before going further, you might like to devise a general description of each of the columns in this array. What are the remainders when you divide each number by 5?

Now look where the values lie which can be reached just using 7p stamps. If you look at $1 \cdot 7, 2 \cdot 7, 3 \cdot 7, 4 \cdot 7$, is it obvious that none of these numbers can lie in the fifth column? (It would contradict the lemma in the comment on **section 9**.) Can you give a reason why no two of them lie in the same column? If they all lie in different columns, and none in the fifth column, then by the time $4 \cdot 7$ is reached every column will have been hit, and so every number from $4 \cdot 7$ onwards is reachable. In fact the last number not to be reached is the number above $4 \cdot 7$, namely $4 \cdot 7 - 5 = (5 - 1) \cdot (7 - 1) - 1$.

If you had started with 6p and 10p stamps instead of 5p and 7p stamps at the very beginning, how would the process have broken down? What are the unreachable values? Suggest other pairs of starting values with a similar defect.

Can the discussion be generalised to deal with stamps with values ap and bp, where $0 < a < b$? How many columns will you use? The critical point is whether the numbers $1 \cdot b, 2 \cdot b, \ldots, (a - 1) \cdot b$ all lie in different columns, and none of them in the ath column. What is the arithmetical condition to ensure this? If this condition is satisfied can you determine the last unreachable value?

We will explore how many unreachable values there are. For each unreachable value u in the original display, examine the number $30 - u$. What do you notice about all the $30 - u$ values? Can you describe the shape formed by the unreachable us and their matching $(30 - u)$s? Can you also describe the symmetry of their relative positions? If an unreachable number $u = 7x + 5y$, with $x = 1, 2, 3$ or 4, what can be said about y? Now $30 - u = 30 - (7x + 5y) = 7(5 - x) + 5(-1 - y)$. If u is unreachable can you prove that $30 - u$ is reachable? How many unreachable us and reachable $(30 - u)$s are there altogether? And what proportion are unreachable? Can you extend this method to the ap and bp case? If so, what is the formula for the number of unreachable numbers?

11 Tests for divisibility

Give rules for testing whether a given number is
  (i)   divisible by 10,
  (ii) divisible by 5,
  (iii) divisible by 2.

If abc is a three figure number (like 247 or 864), is it always divisible by 9 when $a + b + c$ is divisible by 9? Try to justify this rule, and extend it to seven digit numbers.

If abc is a three figure number (like 247 or 864), is it always divisible by 3 when $a + b + c$ is divisible by 3? Try to justify this rule and extend it to seven digit numbers.

If abc is a three figure number (like 247 or 864), can you give a rule or combination of rules to decide whether it is divisible by 6 or not? If you can give a rule, try to justify it, and extend it to many digit numbers.

Is 13 a factor of 247247, of 638638, of any $abcabc$ number? Can you see why? What are the factors of 1001? The number 11 is a factor of 1001. Is 11 a factor of 100001, and 10000001? For the number 7491, alternately add and subtract digits to obtain $7 - 4 + 9 - 1 = 11$. Does this guarantee that 7491 is divisible by 11? Try to justify the extension of this idea to numbers with more digits.

If abc is a three figure number (like 247 or 864), can you give a rule to decide whether it is divisible by 4 or not? Can you give a rule to decide whether it is divisible by 8 or not?

If $abcde$ is a five digit number, show that it is divisible by 7 if and only if $abcd - 2 \cdot e$ is divisible by 7.

Can you modify these rules for work with numbers in different bases?

12 Spot check*

If you were proof reading materials which included some arithmetic, could you decide quickly, without using a calculator, which of the answers here is wrong?

$$19752 + 95234 = 114986 \ or \ 114988?$$

What about $345674 - 19367 = 326303$ or 326307?
What about $3247 \times 2303 = 7477841$ or 7477849?
  Can you construct an addition table giving the last digit of any sum?

* *No comments.*

+	0	1	2	3	4	5	6	7	8	9
0										
1										
2										
3										
4										
5										
6										
7										
8										
9										

Now try to justify the table which we will refer to as the table for $(\mathbf{Z}_{10}, +)$. The basic idea to use is that any positive whole number with last digit 3 (say) must be of the form $10n + 3$; similarly any positive number with last digit 4 must be of the form $10m + 4$. Now $(10n + 3) + (10m + 4) = 10(n + m) + 7$. Make sure you have a sound reason for putting 6 in the $7 + 9$ box.

Can you construct a multiplication table giving the last digit of any product?

$(\mathbf{Z}_{10}, \times)$

×	0	1	2	3	4	5	6	7	8	9
0										
1										
2										
3										
4										
5										
6										
7										
8										
9										

Now justify this table (e.g. the entry $7 \times 9 = 3$) as you did the last one: $(10m + 7)(10n + 9) = 10(10mn + 9m + 7n + 6) + 3$.

13 Eratosthenes' sieve

	1	2	3	4	5	6	7	8	9
10	11	12	13	14	15	16	17	18	19
20	21	22	23	24	25	26	27	28	29
30	31	32	33	34	35	36	37	38	39
40	41	42	43	44	45	46	47	48	49
50	51	52	53	54	55	56	57	58	59
60	61	62	63	64	65	66	67	68	69
70	71	72	73	74	75	76	77	78	79
80	81	82	83	84	85	86	87	88	89
90	91	92	93	94	95	96	97	98	99
100	101	102	103	104	105	106	107	108	109
110	111	112	113	114	115	116	117	118	119
120	121	122	123	124	125	126	127	128	129
130	131	132	133	134	135	136	137	138	139
140	141	142	143	144	145	146	147	148	149
150	151	152	153	154	155	156	157	158	159
160	161	162	163	164	165	166	167	168	169
170	171	172	173	174	175	176	177	178	179
180	181	182	183	184	185	186	187	188	189
190	191	192	193	194	195	196	197	198	199

Working with the numbers 1–99 to start with, encircle the number 2 and strike out all other multiples of 2; encircle the number 3 and strike out all other multiples of 3; encircle the number 5 and strike out all other multiples of 5; encircle the number 7 and strike out all other multiples of 7.

Look at the numbers ($\neq 1$) under 100 which have not been struck out. Are they all prime numbers? Give a reason why every composite number under 100 has been struck off. If $a \cdot b$ is a number less than 100 and neither a nor b is 1, say why either a or b is less than 10. If a composite number has a factor less than 10, must it have a prime factor less than 10?

Why must every composite number less than 200 have a factor less than 15? Deduce that every composite number less than 200 has a factor 2, 3, 5, 7, 11 or 13. In the table above, strike out all the composite numbers up to 200.

What last digit may a prime number (> 5) have?

14 Raffle tickets and neighbours

Avoiding prime raffle tickets

Ann wants to buy three raffle tickets and she feels sure that prime numbers are unlucky, so she does not want any of the numbers to be prime. The raffle tickets are in a book of tickets numbered 1 to 500 and the ticket seller will only allow her to choose adjacent numbers. What number tickets may she choose? Now suppose her friend Bill wants to buy six tickets and again does not want any of the numbers to be primes. Can he find six consecutive non-prime numbered tickets?

Look back at the section on Eratosthenes' sieve. Can you find a run of three, four, five or six consecutive numbers without a prime amongst them? Can you find runs of non-primes of any length?

Look at the following run of length 3:

$$(2 \cdot 3 \cdot 4) + 2 = 26, (2 \cdot 3 \cdot 4) + 3 = 27, (2 \cdot 3 \cdot 4) + 4 = 28$$

Which numbers divide each term? By writing each number as an algebraic expression it is easy to see that 2 divides $(2 \cdot 3 \cdot 4) + 2 = 26$, 3 divides $(2 \cdot 3 \cdot 4) + 3 = 27$ and 4 divides $(2 \cdot 3 \cdot 4) + 4 = 28$.

Try the run $(2 \cdot 3 \cdot 4 \cdot 5) + 2, \ldots, (2 \cdot 3 \cdot 4 \cdot 5) + 5$. Again which numbers divide each term? Find a run of six consecutive numbers each of which is divisible by $2, \ldots, 7$ respectively.

Now you should be able to give a general construction of a sequence of any finite length with the first term having a divisor of 2, the next a divisor of 3 and so on. That is, we have shown that there exist arbitrarily long sequences of consecutive non-primes.

Neighbours

Mary was chatting to her next door neighbour James outside their front doors and Mary said "Hey look!", pointing at the two door numbers, "That's interesting; our house numbers are both primes." James then noticed that his other neighbour's door number was also a prime. What is the number of James's house? Is there really only one possible answer?

Apart from the obvious pair 2,3, all other pairs of primes have at least one even number between them. When two prime numbers differ by 2 (like 17 and 19) we will call them a "prime pair". Are there any other prime pairs? Find all prime pairs less than 200 using the table constructed with Eratosthenes' sieve.

There are several interesting properties of prime pairs. Consider the sum of prime pairs greater than the pair 3,5. What common property do all these sums have? Work out the product of a prime pair plus 1. What do you notice? Will this always be true? To return to the question we started with. Can you find any prime triples; that is, numbers p, $p + 2$ and $p + 4$, all of which are primes? Will there be any others?

15 How many primes?

Work out the number $2 \cdot 3 \cdot 5 \cdot 7 \cdot 11 \cdot 13 + 1$ using a calculator.

Might 2 be a factor of this number? Or 3? Or 5? Or 7? Or 11? Or 13? What is the remainder when this number is divided by each of these primes?

If 30031 is not prime, say why it must have a prime factor between 13 and 173. Try to find its factors.

If $p_1, p_2, \ldots, p_n$ are the first n prime numbers, construct a number which is not divisible by any of them.

If someone claimed that there might only be a finite number of primes, prove them wrong.

16 No shuffling

Wendy, Toots, Tweedle and I play bridge every Wednesday night. On the Wednesday I want to tell you about, I was to partner Toots, and Toots and I sat North/South. Tweedle and Wendy sat East/West and I was the dealer.

I had got an unusual pack containing 100 cards. It was all of one suit and contained neither aces nor picture cards. When I got the cards, they were numbered in order from 1 to 100, and I didn't shuffle them. So when I began dealing, Wendy got the card numbered 1, Toots the card numbered 2, Tweedle the card numbered 3 and so on. When we looked at the 25 cards we had each been dealt, the others indignantly asked how I expected them to play bridge with these, adding "*** *** ***!", but then they began to get interested in the hands they had been dealt.

What cards had I been dealt? Is there a simple description?

What cards had Wendy been dealt? What is the remainder when you divide each of Wendy's numbers by 4? What cards had Toots and Tweedle been dealt?

If Wendy puts down two cards, with numbers less than 50 on them, who holds the card with the sum of those numbers on it? What about other sums?

Now try possible products, but beware of size (or else pretend that the pack of cards is bigger). Who holds the cards with square numbers on them? Who holds prime numbers? Who holds many primes?

17 Can you really tell the time?

One of my clocks has lost its hour hand and when it is on the other side of the room I can't read the numbers on its face. But I can still see whether the minute hand is pointing upwards, downwards or to the left or right. So I can use it for the quarter-hours.

The numbers in the table below are numbers of quarter-hours. Positive numbers measure time into the future. Negative numbers measure time into the past.

. . .	. . .	. . .	. . .
−20	−19	−18	−17
−16	−15	−14	−13
−12	−11	−10	−9
− 8	−7	−6	−5
−4	−3	−2	−1
0	1	2	3
4	5	6	7
8	9	10	11
12	13	14	15
16	17	18	19
20	21	22	23
. . .	. . .	. . .	. . .

The number 0 marks time present, with the minute hand pointing straight up-wards. At what numbers of quarter-hours, past and future, will the minute hand be pointing vertically upwards, on the hour? At what numbers of quarter-hours will the minute hand be pointing to the right at the first quarter? At what numbers of quarter-hours will the minute hand be pointing downwards at the half-hour? At what numbers of quarter-hours will the minute hand be pointing to the left at the third quarter?

Which column contains +100?
Which column contains −100?
Which column contains +99?
Which column contains −99?

If two numbers of quarter-hours lie in the same column, what can you say about the amount of time which separates them? The mathematical convention describing two numbers in the same column here is to say that the two numbers are *congruent modulo 4*. So 18 and −10 are *congruent modulo 4*, and this is *also* written symbolically as $18 \equiv -10 \pmod 4$. (Three of these are true, three are false: $6 \equiv -6 \pmod 4$, $23 \equiv -23 \pmod 4$, $300 \equiv 500 \pmod 4$, $17 \equiv 5 \pmod 4$, $17 \equiv -5 \pmod 4$, $16 \equiv -18 \pmod 4$.)

Is every whole number congruent to 0, 1, 2, or 3 *modulo 4*?

If the minute hand of the clock stands at	and time moves on through a further				
	0	1	2	3	quarter-hours
	then the minute hand is at				
the hour	the hour	quarter-past	half-past	quarter-to	
quarter-past	quarter-past	half-past	quarter-to	the hour	
half-past	half-past	quarter-to	the hour	quarter-past	
quarter-to	quarter-to	the hour	quarter-past	half-past	

This can be summarised rather neatly in the form

	0	1	2	3
0	0	1	2	3
1	1	2	3	0
2	2	3	0	1
3	3	0	1	2

where the numbers represent whole columns of integers in the table at the top; that is, positions of the minute hand. The combinations describe the addition of time.

Another of my clocks has lost its minute hand. Fortunately the numbers on its face stand out clearly. Numbers in this table are numbers of hours. Positive numbers measure time into the future. Negative numbers measure time into the past.

$\cdots$	$\cdots$	$\cdots$	$\cdots$	$\cdots$	$\cdots$	$\cdots$	$\cdots$	$\cdots$	$\cdots$	$\cdots$	$\cdots$
−36	−35	−34	−33	−32	−31	−30	−29	−28	−27	−26	−25
−24	−23	−22	−21	−20	−19	−18	−17	−16	−15	−14	−13
−12	−11	−10	−9	−8	−7	−6	−5	−4	−3	−2	−1
0	1	2	3	4	5	6	7	8	9	10	11
12	13	14	15	16	17	18	19	20	21	22	23
24	25	26	27	28	29	30	31	32	33	34	35
$\cdots$	$\cdots$	$\cdots$	$\cdots$	$\cdots$	$\cdots$	$\cdots$	$\cdots$	$\cdots$	$\cdots$	$\cdots$	$\cdots$

How far must time move on before the hour hand is again in the same position? Do the columns in the array correspond to positions of the hour hand?

Two numbers which lie in the same column are said to be congruent *modulo* 12, so 14 and −10 are congruent *modulo* 12 and this is written symbolically $14 \equiv -10 \pmod{12}$. (Three of these are true, three are false: $9 \equiv 3 \pmod{12}$, $9 \equiv -3 \pmod{12}$, $100 \equiv 4 \pmod{12}$, $100 \equiv -4 \pmod{12}$, $45 \equiv 7 \pmod{12}$, $87 \equiv 15 \pmod{12}$.)

Fill in this table for hours shown on a clock face.

+	0	1	2	3	4	5	6	7	8	9	10	11
0												
1												
2												
3												
4												
5												
6												
7												
8												
9												
10												
11												

As before, the numbers represent columns and there is one number here from each column. Would the addition have worked out the same if another representative of a column had been used?

Can you give an algebraic rule to test whether $a \equiv b \pmod{4}$? Can you give an algebraic rule to test whether $a \equiv b \pmod{12}$?

18 Fibonacci numbers and the division algorithm

We have seen in **7. Breeding rabbits** that the highest common factor of two consecutive terms in the Fibonacci sequence always seems to be 1. We can use the division algorithm given in **5. Catching practice** to establish this result.

Let us begin by proving that the highest common factor of $f(6) = 8$ and $f(5) = 5$ is 1. We first express $f(6)$ as a multiple of $f(5)$ plus a remainder:

$$8 = 1 \cdot 5 + 3$$

Any divisor of 8 and 5 must also be a divisor of 3. For if n divides a and b and $a = b + c$ then n also divides c.

Now express 5 as a multiple of 3 plus a new remainder:

$$5 = 1 \cdot 3 + 2$$

Any divisor of 5 and 3 must also be a divisor of 2.

We continue with this process until our remainder is 0 and then declare the last non-zero remainder to be the highest common factor:

$$3 = 1 \cdot 2 + 1$$
$$2 = 2 \cdot 1 + 0$$

The last non-zero remainder is 1. Hence the highest common factor of $f(6)$ and $f(5)$ is 1.

Do the same with $f(7)$ and $f(6)$.

Notice how many applications of the division algorithm are needed each time. For $f(6)$ and $f(5)$ we needed four applications of the division algorithm to obtain a remainder of 0. How many applications are needed for $f(7)$ and $f(6)$? How many applications do you think it will take to establish that the highest common factor of $f(8)$ and $f(7)$ is 1?

Can you give two numbers such that it will take 10 applications of the division algorithm to find that their highest common factor is 1?

For any number n we can always find two numbers that will take n applications of the division algorithm to find their highest common factor, for example $f(n + 2)$ and $f(n + 1)$. Thus we can see that there is no upper bound on the number of applications needed by the the division algorithm to find the highest common factor of any two numbers.

19 Relations can be difficult

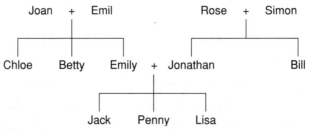

Figure 3

Symmetric relationships

From the family tree you can see that Penny is Lisa's sister and Lisa is Penny's sister. Is it always true that if A is B's sister is B always A's sister? Can you find someone's name to go in place of X?

Lisa is X's sister but X is not Lisa's sister.

For the relationship "is a sibling of" (a sibling is either a brother or a sister) does this relationship always work both ways?

Relationships such as "is a sibling of" are said to be *symmetric* because

If A is a sibling of B then B is a sibling of A for all possible A and B.

In the definition of a symmetric relationship it is not enough to find one example where the relationship works in both ways. A relationship is only called a symmetric relationship if the symmetry holds in all cases. If A is a sister of B then B may well be a sister of A, but this is not always the case, so "is a sister of" is *not* a symmetric relationship in most families.

Which of the following relationships are symmetric?

▷   has the same parents as
▷   is the father of
▷   is married to
▷   is an aunt of

We can think about relations other than those between people. For example, we can talk about relations between numbers. Let x and y be integers; then

x is greater than y

is a relation and we can ask: does this imply that y is greater than x? Obviously "is greater than" is *not* a symmetric relation.

Our formal definition of a symmetric relation is therefore: a relation **R** on a set S is said to be symmetric if

$$a\mathbf{R}b \Rightarrow b\mathbf{R}a \text{ for } a,b \text{ in } S$$

Here are some more relations and objects that they compare. Decide which of them are symmetric. For those that are not symmetric give a counter-example.

1. "is less than" for integers.
2. "is the opposite of" for words in English.
3. "is the same distance from the origin as" for points on a plane.
4. "is a factor of" for integers.
5. "is congruent *modulo* 4 to" for integers.
6. "is perpendicular to" for lines in a plane.
7. "is parallel to" for lines in a plane.

Transitive relationships

To return to our family tree, can you find A, B, C so that A is the same sex as B, B is the same sex as C and A is the same sex as C? Does the third "is the same sex as" follow automatically from the first two?

Can you also find three people so that A is the son of B and B is the son of C? Does it then follow that A will be the son of C?

We say that "is the same sex as" is a *transitive* relationship and "is the son of" is *not* a transitive relationship. A relation **R** on a set S is said to be transitive if

$$a\mathbf{R}b \text{ and } b\mathbf{R}c \Rightarrow a\mathbf{R}c \text{ for } a,b,c \text{ in } S$$

For the examples of relations given above in 1 to 7 decide which ones are transitive.

Reflexive relationships

Here is a third property that a relationship may or may not have. For example "is the same sex as". Is it true that Lucy is the same sex as Lucy? Is it true for any A that A is the same sex as A? A relationship with this property is called *reflexive*.

Is the relationship "is the mother of" reflexive?

As before we can apply the same ideas to objects as well as people. For example, the relation "is less than or equal to" for integers is reflexive since any number is less than or equal to itself.

A relation **R** on a set S is said to be reflexive if

$$a\mathbf{R}a \text{ for all elements } a \text{ in } S$$

For the relations given above in 1 to 7 decide which ones are reflexive.

The relation "is congruent *modulo* 4 to" has all three properties. We can consider the numbers between 0 and 10 and see which ones are related. For example, for the numbers 0, 4 and 8, $0 \equiv 4$, $0 \equiv 8$ and $4 \equiv 8$. Choose a number other than 0, 4 or 8 and find another set of numbers which are all related to each other under this relation. Keep doing this until all the numbers up to 10 have been considered.

You will have found that the relation has partitioned the set into classes $\{0,4,8\}$, $\{1,5,9\}$, $\{2,6,10\}$ and $\{3,7\}$. In each class everything is related to everything else in the class and not to any element of another class.

Using the same set, 0 to 10, see how the relation $a \equiv b \pmod{5}$ divides up the set into classes. Are the classes again distinct?

For the relation "is congruent *modulo* n to" try to *prove* that it has all three properties.

Non-transitive dice

Suppose that you have three dice. Die *A* has the numbers 6, 6, 2, 2, 2, 2 on its six sides. Die *B* has the numbers 5, 5, 5, 5, 1, 1. Die *C* has the numbers 4, 4, 4, 3, 3, 3.

You challenge a friend to the following game. Your friend first chooses any die, after which you choose one of the two remaining dice. Each of you then throws your die and the person with the higher score wins. Is the game fair?

After a little experimentation you should be able to see that if you played this game for a long time, die A would beat die B more than half the time and die B would beat die C more than half the time, but die C would beat die A more than half the time. This surprising result is a consequence of the fact that the relation "beats" on the set of three dice is not transitive.

For the dice A and B, draw up a table with six rows and six columns, showing which die wins for each possible combination of results. Count the number of wins for A and the number for B. Do the same for dice A and C, and for dice B and C. What should be your strategy for ensuring that you have an unfair advantage?

20 Dominoes

Imagine standing a set of dominoes on their ends (like toy soldiers). Each domino is 5 cm from its neighbour and in front of the first domino is a pea shooter. I fire the pea shooter at the first domino. Will the whole line of dominoes fall over? There are two questions we need to ask: does the pea knock the first domino over, and is the distance apart of the dominoes small enough that each domino falling will knock over its neighbour?

A very useful method of proof depends on just two such ideas, namely (i) can you reach the first step; and (ii) can you get from any step to the next? These ideas are basic to the possibility of counting. Here are some statements, and some evidence to support them.

1. $1^2 + 2^2 + 3^2 + \ldots + n^2 = \frac{1}{6}n(n+1)(2n+1)$. True for $n = 1$, $n = 2$ and $n = 3$.
2. $n^2 \geq 2^{n-1}$. True for $n = 2$, $n = 3$ and $n = 4$.
3. $1 + 2 + 3 + \ldots + n = \frac{1}{2}n(n+1)$. True for $n = 1$, $n = 2$ and $n = 3$.

Which statements are true for all larger values of n? How many values of n do we need to check?

Consider example 1. Suppose we know that for some particular value of n,

$$1^2 + 2^2 + 3^2 + \ldots + n^2 = \frac{1}{6}n(n+1)(2n+1) \tag{20.1}$$

then what can we say about $1^2 + 2^2 + 3^2 + \ldots + n^2 + (n+1)^2$?

We can try to find out by adding $(n+1)^2$ to both sides of our equation (20.1). By rearranging $\frac{1}{6}n(n+1)(2n+1) + (n+1)^2$, show that equation (20.1) is true with n replaced by $n + 1$.

So now we have shown that if the statement holds for n then it will hold for $n + 1$. Hence, since it holds for 3 it holds for 4 and since it holds for 4 it holds for 5 and so on.

Try to use the same method on example 3.

The principle used here is called **the principle of mathematical induction**:

Let $S(n)$ be a statement that is either true or false for each positive integer n. Suppose that

(i) $S(1)$ is true and
(ii)   for all positive integers n, $S(n+1)$ is true whenever $S(n)$ is true;

then $S(n)$ is true for all $n \geq 1$.

Condition (i) is called the basis for the induction and condition (ii) is called the inductive step.

Here are three more examples for you to prove using the principle of mathematical induction:

4. $1^3 + 2^3 + 3^3 + \ldots + n^3 = [(\frac{1}{2})n(n+1)]^2$ for $n \geq 1$.
5. $6^n - 5n + 4$ is divisible by 5 for $n \geq 1$.
6. $1 + a + a^2 + a^3 + a^4 + \ldots + a^n = (a^{n+1} - 1)/(a-1)$ for $n \geq 1$.

Do we really need to prove **both** the basis and the induction step?

Both parts of the induction are important as can be seen in the following examples:

7.   Show that the statement $2 + 4 + \ldots + 2n = (n+2)(n-1)$ for all $n \geq 1$ satisfies the inductive step but has no basis.
8.   Show that for some values of n, $n^2 + n + 41$ is a prime number and for some it is not, so that there is no inductive step which would show that $n^2 + n + 41$ is a prime number for all possible n.

Induction on Fibonacci numbers

Fibonacci numbers, numbers in the sequence 1, 1, 2, 3, 5, 8, . . . , were introduced in **section 7**. Can you write the numbers 1 to 10 as the sums of distinct Fibonacci numbers? For example,

$$6 = 5 + 1$$

Can 100 be expressed in this way? Can *every* positive integer be expressed as a sum of distinct Fibonacci numbers? Try to give a proof that all numbers can be expressed in this way.

Subsequences

$$2, 8, 34, 144, 610, 2584, 10946, 46368, 196418, 832040, \ldots$$

This sequence is obtained by taking every third term of the Fibonacci sequence.

If we form a new sequence by taking just some of the terms of a sequence we call this new sequence a *subsequence* of the original sequence. For example, we can describe the above subsequence as $A(n)$ where $A(n) = f(3n)$.

These first few terms $A(1)$, $A(2)$, . . . , $A(10)$ are all even numbers. Must this be true for all $A(n)$?

We can rewrite $A(2)$ as

$$\begin{aligned}
A(2) = 8 = 5 + 3 &= f(5) + f(4) \\
&= [f(4) + f(3)] + f(4) \\
&= 2f(4) + f(3) \\
&= 2{\cdot}3 + A(1)
\end{aligned}$$

$$\begin{aligned}
A(3) = 34 = 21 + 13 &= f(8) + f(7) \\
&= [f(7) + f(6)] + f(7) \\
&= 2f(7) + f(6) \\
&= 2{\cdot}13 + A(2)
\end{aligned}$$

Find $A(n+1) - A(n)$.

Now prove that every term of $A(n)$ will be divisible by 2.

Write down the first six terms of the sequence $B(n) = f(4n)$ obtained by taking every fourth term. Is there a common factor amongst these terms? Write down an expression for $B(2)$, $B(3)$ and for $B(n+1) - B(n)$. Convince yourself that this can be used to prove that every fourth term is always divisible by 3.

Is there a similar subsequence of terms all divisible by 5?

PART II

21 Chinese remainders

If an integer is congruent to 9 *modulo* 15, you should be able to claim that it is congruent to 0 *modulo* 3 and congruent to 4 *modulo* 5. This section is about the converse result.

In a Chinese book well over 1000 years old (perhaps 2000), called *Master Sūn's Mathematical Manual*, the following problem is posed:

> There is an unknown number of things.
> Three by three, two remain; five by five, three remain;
> seven by seven, two remain. How many things?

This means that if the unknown number is divided by 3, the remainder is 2; if the unknown number is divided by 5, the remainder is 3; and if the unknown number is divided by 7, the remainder is 2. What can the number be? This kind of problem may have arisen in calculating the calendar or as a secretive way of giving information about numbers of soldiers.

The problem amounts to finding a number which lies in the third column of the first array, in the fourth column of the second array and in the third column of the third array, shown here:

0	1	2		0	1	2	3	4		0	1	2	3	4	5	6
3	4	5		5	6	7	8	9		7	8	9	10	11	12	13
6	7	8		10	11	12	13	14		14	15	16	17	18	19	20
9	10	11		15	16	17	18	19		21	22	23	24	25	26	27
...	...	...		...	...	...	...	...		...	...	...	...	...	...	...

The three conditions make the problem particularly hard to think about, but we can examine the first two conditions (dividing by 3 gives remainder 2 and dividing by 5 gives remainder 3) by examining each of the numbers 1, 2, 3, ..., 15 and entering them in the appropriate square of the following table.

Remainder on dividing by 5

	0	1	2	3	4
Remainder 0					
on dividing 1					
by 3 2					

If you have filled the squares in carefully you will have put one number in each square, giving you just one number up to 15 which satisfies both the conditions. Examine the table until you can detect a simple routine for entering the numbers in it.

You should have found that if a number has remainder 2 when divided by 3, and remainder 3 when divided by 5, then it has remainder 8 when divided by 15.

Now we can tackle the full problem, by enterin⁰g the numbers from 1 to 105 in the appropriate square of the following table.

Remainder on dividing by 15

		0	1	2	3	4	5	6	7	8	9	10	11	12	13	14
	0															
Remainder	1															
on	2															
dividing	3															
by	4															
7	5															
	6															

If you have entered the numbers carefully (and there is a simple routine for doing this) you will have put just one number in each square, and therefore found just one number under 105 which has remainder 2 when divided by 7 and remainder 8 when divided by 15. Check that this number satisfies all three original conditions.

Keep the divisors 3, 5 and 7 but alter the remainders in the original problem, and determine the solution of the problem you have formulated.

Had the original problem said "Four by four, two remain; six by six, four remain", what might have been the number of things?

Had the original problem said "Four by four, one remains; six by six, two remain", what might have been the number of things?

Explain why the divisors {4, 6} give such different results to the divisors {3, 5, 7}.

22 Systematic catching practice

When there are 12 cricketers in a circle throwing the ball around the circle with a regular number of people missed out with each throw, steps of 1, 5, 7 and 11 reach every person, steps of 2, 3, 4, 6, 8, 9 and 10 do not. When there are nine in the circle, steps of 1, 2, 4, 5, 7 and 8 reach every person, steps of 3 and 6 do not. When there are 11 in the circle, steps of all sizes from 1 to 10 reach every person.

If with a step of length l every person is reached around a circle of size 12, then the highest common factor of l and 12 is . . . ? If with a step of length l every person is reached around a circle of size n, the highest common factor of l and n is . . . ?

Euler's ϕ function (sometimes called the *totient* function) counts how many step lengths reach every position around a circle, so $\phi(12) = 4$ because the set {1, 5, 7, 11} has four numbers in it. From the work you have done, write down $\phi(9)$ and $\phi(11)$.

What is $\phi(5)$? By writing the numbers from 1 to 25 in five rows and five columns, see why $\phi(25) = 20$. By writing the numbers from 1 to 125 in 25 rows and five columns, see why $\phi(125) = 100$.

Can you generalise this so that for any prime number p you can write down $\phi(p)$, $\phi(p^2)$, $\phi(p^3)$, $\phi(p^4)$, etc.?

23 Do you know your tables?

Look back to **12. Spot check** for a table giving the last digits in a multiplication table. Each number in the table stands for an infinity of numbers, with a standing for any number of the form $10k + a$. How did you show that ab has the same last digit as $(10k + a)(10l + b)$?

Can you show that if $x \equiv a$ (mod 10) and $y \equiv b$ (mod 10), then $xy \equiv ab$ (mod 10)?

Look back to **16. No shuffling** for a multiplication table of people. Each name stood for the numbers of the cards in the person's hand. Wendy's were all of the form $4k + 1$, Toots' were all of the form $4k + 2$, Tweedle's were of the form $4k + 3$ and mine were all of the form $4k$. How can you express the connection between two cards in the same player's hand with mathematical symbols?

Can you show that if $x \equiv a$ (mod 4) and $y \equiv b$ (mod 4), then $xy \equiv ab$ (mod 4)? If you are not sure, take a and b as particular numbers, 1 and 2 say.

Show that if $x \equiv a$ (mod 3) and $y \equiv b$ (mod 3), then $xy \equiv ab$ (mod 3); then fill in the multiplication table *modulo* 3:

.	0	1	2
0			
1			
2			

Your preliminary argument allows the number a to stand for any number of the form $3k + a$.

Show that if $x \equiv a$ (mod n) and $y \equiv b$ (mod n), then $xy \equiv ab$ (mod n), and this theorem justifies a multiplication table for any modulus.

Make multiplication tables for multiplication *modulo* 4, 5, 6, 7, 8, 9, 10, 11 and 12. Is there, in each table, a number I, such that $xI \equiv Ix \equiv x$ (mod n) for all values of x? Such a number is called an *identity* for the table.

When $xy \equiv I$ (mod n), where I is the identity *modulo* n, x and y are said to be *inverses* of each other. Look at the tables you have made for multiplication, and decide which numbers have inverses and which do not.

In your multiplication tables, some of the rows contain all the numbers in the leading row outside the top of the table, and some do not. Can you distinguish the rows which contain all the numbers from those which do not?

Can you link your answers to the last two paragraphs to the list of generators for the addition tables in **22. Systematic catching practice**?

24 Coding and decoding

What do you think the following says?

GPIZIV HMGO

The easiest codes to crack are the ones where every letter in our message is replaced by the corresponding letter of the alphabet the same number of places on. In this example we have moved the letters four places forwards to encode. If we represent each letter by an integer, A by 0, B by 1 and so on, then we can give the code with modular arithmetic.

Add 4 to the number of the letter and reduce *modulo* 26 if necessary. For example, A has number 0, 0 + 4 = 4 and 4 corresponds to the letter E. X has number 24, 24 + 4 = 28 which reduces to 2 *modulo* 26 and is the letter B.

number to number + 4 (*modulo* 26)

Original letter	ABC	DEF	GHI	J KL	MNO	PQR	ST U	VWX	Y Z
Encoded letter	EFG	HI J	KLM	NOP	QR S	TUV	WXY	ZA B	CD

This is an example from the family of shift transformations of the form:

number to number + a constant integer (*modulo* some number):

$$x \text{ to } x + c \pmod{26}$$

An early user of such a code was Julius Caesar and hence these codes are also known as Caesar codes. Can you discover the shift transformation used in the following Caesar code?

HW WX EUXWH

In a Caesar code the code breaker only has to check 26 possible shift transformations so it is not a very safe code. A variation on this is to use the transformation

$$x \text{ to } px + q \pmod{26}$$

In order to see how much more secure this code is, we need to see how many possible codes there are of this form. Explore what happens if we take $p = 2$ and $q = 0$. Which values for p are going to give 26 different values for $px \pmod{26}$ as x runs from 0 to 25? So how many codes are there replacing x by $px \pmod{26}$? And how many are there replacing x by $px + q$? For a computer this is not a large number and hence again none of these codes are very secure.

A better method is to permute the letters randomly. This will give us 26! different codes. Here is another code for you to try your skill on:

ACNH NH Q HYZOYA PYHHQLY

This time we must use some of our knowledge of the English language. Which letters can occur on their own as a word, which letters occur as doubles and what are the most frequently occurring letters?

The game of Scrabble has been designed so that frequently occurring letters in our language have a low point score and rarely occurring ones have a high point score. The points for using the different letters are as follows:

Points	Letters
1	A E I L N O R S T U
2	D G
3	B C M P
4	F H V W Y
5	K
8	J X
10	Q Z

Try to use this information to crack the coded message above.

25 Repacking

A tennis ball manufacturer always packages tennis balls into square trays for ease of distribution. The trays are all one tennis ball deep but the other two sides are equal but of variable length. For example, one of the trays is $4 \times 4 \times 1$. The sports shop owner knows that her customers will not want to buy the balls in these large quantities and so repackages the balls in tubes. The tubes used to repack the balls hold five tennis balls each. After repacking a whole tray she puts any balls left to one side. The shop owner never knows what size of square tray will arrive but has found that no matter what size the square tray was she never has two balls left over.

By listing all the possible square numbers 1, 4, 9, . . . between 1 and 50 in the table below decide what possible numbers of balls could be left over from a square tray, containing at most 50 balls, repackaged into tubes of size 5.

Balls left over

No balls	One ball	Two balls	Three balls	Four balls
				3^2

Have you found any square tray sizes that would result in two balls left over after repackaging? If the shop owner received larger square trays of tennis balls and still packed them into tubes of fives might she sometimes have two left over?

A different sports shop repackages the square tray into cartons of seven balls. What are the possible numbers of left over balls in this case?

26 Where have all the squares gone?

0	1	2
3	4	5
6	7	8
9	10	11
12	13	14
15	16	17
18	19	20
21	22	23
24	25	26

0	1	2	3
4	5	6	7
8	9	10	11
12	13	14	15
16	17	18	19
20	21	22	23
24	25	26	27
28	29	30	31
32	33	34	35

0	1	2	3	4
5	6	7	8	9
10	11	12	13	14
15	16	17	18	19
20	21	22	23	24
25	26	27	28	29
30	31	32	33	34
35	36	37	38	39
40	41	42	43	44

In the three displays above, circle all the square numbers. Are there columns which do not contain any square numbers? Do you think this is just because we have not looked very far up the sequence of numbers, or do you think this will be true how-ever far the displays are extended?

A way to investigate this question for the *first* display (the one with three columns in it) is to start from the fact that every whole number takes one of the forms $3n$, $3n + 1$, $3n + 2$ for some n, because, when divided by 3, each number leaves remainder 0, 1 or 2. If we square each of these three possibilities we can decide which of the three columns of the original array contains the result:

$(3n)^2 = 9n^2 = 3(3n^2)$, in the first column.
$(3n + 1)^2 = 9n^2 + 6n + 1 = 3(3n^2 + 2n) + 1$, in the second column.
$(3n + 2)^2 = 9n^2 + 12n + 4 = 3(3n^2 + 4n + 1) + 1$, in the second column.

Since these three types exhaust the possibilities, no squares can ever appear in the third column.

You can tackle the location of squares in the columns of the *second* and *third* displays in a similar fashion.

27 Where have all the squares come from?

If the positive integers are displayed in five columns with 0, 1, 2, 3 and 4 in the first row, the columns in which the first few square numbers appear are as follows:

0	1	2	3	4
0	1			4
25	16			9
100	36			49
225	81			64
	121			144
	196			169

In the previous section we had already worked out that all the squares appeared in the columns with 0, 1 and 4 at the head. Now take each of these columns in turn, and for each square number x^2 in that column, find out in which columns x itself might have been.

If the positive integers are displayed in seven columns with 0, 1, 2, 3, 4, 5 and 6 in the first row, the columns in which the first few square numbers appear are as follows:

0	1	2	3	4	5	6
0	1	9		4		
49	36	16		25		
196	64	100		81		
	169	121		144		
	225	289		256		

Check that these square numbers have been located in the appropriate columns. Now take each of these columns in turn, and for each square number x^2 in that column, find out in which columns x itself might have been.

If the positive integers were displayed in p (a prime number greater than 2) columns with 0, 1, 2, . . . , x, . . . , y and $p - 1$ in the first row, and x^2 and y^2 appeared in the same column, but not under 0, conjecture a relationship between x and y.

0	1	. . .	x	. . .	y	. . .	$p - 1$
			x^2				
			y^2				

28 How old is Grandma?

One afternoon Amelia was talking to her Grandma about her ambitions to follow in her footsteps and become a mathematics professor. "How old are you Grandma?" she asked. Grandma thought for a while and replied, "If I square the number of papers I wrote and subtract the square of my age then the answer is 119143." "That is not fair", Amelia cried, "I have no idea how many hundreds of papers you wrote". "But I thought you wanted to be a mathematician? Let us see if you can solve it with a few hints from your old Grandma."

Let h be the number of papers and g be Grandma's age. Express Grandma's age in terms of 119143 and h.

So now we want a number for h whose square minus 119143 leaves a perfect square. What is the smallest number worth trying for h?

That still leaves a lot of h numbers to check whether $h^2 - 119143$ is a square. Is there an easy way to tell that $573 = 346^2 - 119143$ is not a square? By squaring all the numbers between 1 and 10 discover which numbers can be the last digit of a square number. (Look at the numbers down the diagonal of the multiplication table in **12. Spot check**).

h		1	2	3	4	5	6	7	8	9	10
h^2											
$h^2 - 3 \equiv g^2$ mod 10											

How can we also use the information in the row labelled $h^2 - 3$ to reduce the number of values of h that we have to check? What values are possible for the last digit of h? Write down the four smallest numbers that might be worth trying for h. Use a calculator to test each one. Now what might Grandma's age be?

Can you see how to use the values of g and h that you have just obtained to factorise 119143? Now you can be sure that only one value of g and of h can satisfy $h^2 - g^2 = 119143$ because the factors of 119143 are prime and thus unique.

Try to use the same method to factorise 2279. What sort of numbers can be factorised by this method?

This is known as **Fermat's factorisation method** and is particularly useful in searching for factors of numbers which are the product of two primes close to the square root of the number. Fermat described his technique for factoring large numbers in a letter believed to have been written to the monk Marin Mersenne in 1643.

What is the prime factorisation of 156379? We will look for h and g such that $h^2 - 156379 = g^2$.

In order to reduce further the numbers we need to check as possible values of h, we can discover which digits are the possible last *two* digits of a square. We do not need to work out all the squares from 1 to 100. The squares from 1 to 25 tell us all the possibilities.

By writing 52 as $50 + 2$ we can see that

$$52^2 = (50 + 2)^2 = 2500 + 200 + 4 \equiv 4 \equiv 2^2 \text{ (mod 100)}$$
$$48^2 = (50 - 2)^2 = 2500 - 200 + 4 \equiv 4 \equiv 2^2 \text{ (mod 100)}$$
$$98^2 = (100 - 2)^2 = 10000 - 400 + 4 \equiv 4 \equiv 2^2 \text{ (mod 100)}$$

Is $1^2 \equiv 49^2 \equiv 51^2 \equiv 99^2$ (mod 100)? Is $3^2 \equiv 47^2 \equiv 53^2 \equiv 97^2$ (mod 100)?

Complete the table below and use the results to obtain the prime factors of 156379 remembering that $h^2 \equiv (50 - h)^2 \equiv (50 + h)^2 \equiv (100 - h)^2$ (mod 100).

h	0	1	2	3	4	5	6	7	8	9	10	11	12	13	14	15
h^2	00	01	04	09	16	25	36	49	64	81	00	21	44	69	96	25
g^2																

h	16	17	18	19	20	21	22	23	24	25
h^2	56	89	24	61	00	41	84	29	76	25
g^2										

29 Higher powers

If the positive integers are displayed in five columns with 0, 1, 2, 3 and 4 in the first row, the columns in which the first few square numbers appear are as follows:

0	1	2	3	4
0	1			4
25	16			9
100	36			49
225	81			64
	121			144
	196			169

To look for fourth powers, we can square and square again. If we start from any number under one of the headings 1, 2, 3 and 4, where may its square lie? Where then may the square of the square lie? So where do fourth powers lie?

If the positive integers are displayed in seven columns with 0, 1, 2, 3, 4, 5 and 6 in the first row, the columns in which the first few square numbers appear are as follows:

0	1	2	3	4	5	6
0	1	9		4		
49	36	16		25		
	64	100		81		
	169	121		144		
	225	289		256		

The columns in which the first few cubes lie are as follows:

0	1	2	3	4	5	6
0	1					27
343	8					125
2744	64					216
	512					1000
	729					1728
	1331					2197

By examining numbers of the type $(7m + a)^3$ for $a = 0, 1, 2, 3, 4, 5$ and 6, or by using the tables given in the comments on **section 23**, deduce that the distribution of cubes shown above continues indefinitely downward. Sixth powers are both squares and cubes. If you start from one of the numbers below the headings 1, 2, 3, 4, 5 or 6 where do the sixth powers lie?

If the positive integers are displayed in 13 columns with 0, 1, 2, 3, 4, 5, 6, 7, 8, 9, 10, 11 and 12 in the first row, the columns in which the first few square numbers appear are as follows:

0	1	2	3	4	5	6	7	8	9	10	11	12
0	1		16	4					9	36		25
169	144		81	121					100	49		64
			196...									

Are these the only columns in which squares may appear? Find the source columns for each of the columns containing squares. Determine the columns in which fourth powers may lie.

The columns in which the first few cubes lie are as follows:

0	1	2	3	4	5	6	7	8	9	10	11	12
0	1				343			8				64
2197	27				512			125				1000
	729				1331			216				1728

Are these the only columns in which cubes may appear? Which columns may contain both cubes and fourth powers? Which columns may contain twelfth powers?

30 Just shuffling and then power (*modulo* 7)

Every integer is congruent to exactly one of 0, 1, 2, 3, 4, 5 or 6 *modulo* 7. Use a calculator to find 2^6, 3^6, 4^6, 5^6, 6^6 and 7^6.

For each of these sixth powers find the remainder on division by 7 to see to which of the integers between 0 and 6 these powers are congruent *modulo* 7. Knowing that $100 \equiv 2 \pmod 7$, can you decide, without using a calculator, to which of the integers between 0 and 6, 100^6 is congruent *modulo* 7. What about 1000^6?

x	$=$	0	1	2	3	4	5	6
$0 \cdot x$	$\equiv$							
$1 \cdot x$	$\equiv$							
$2 \cdot x$	$\equiv$							
$3 \cdot x$	$\equiv$				5			
$4 \cdot x$	$\equiv$							
$5 \cdot x$	$\equiv$							
$6 \cdot x$	$\equiv$							

Fill in each row of this multiplication table using only the integers 0, 1, 2, 3, 4, 5 and 6 in such a way as to enter the integer congruent to the product to be calculated *modulo* 7. The isolated 5 appears because $3 \cdot 4 \equiv 5 \pmod 7$.

When you worked out the entries for 2·1, 2·2, 2·3, 2·4, 2·5 or 2·6 you should have found the integers 1, 2, 3, 4, 5 and 6 just shuffled about. So then

$$(2{\cdot}1){\cdot}(2{\cdot}2){\cdot}(2{\cdot}3){\cdot}(2{\cdot}4){\cdot}(2{\cdot}5){\cdot}(2{\cdot}6) \equiv 1{\cdot}2{\cdot}3{\cdot}4{\cdot}5{\cdot}6 \pmod 7$$

and by rearranging the terms

$$2^6{\cdot}(1{\cdot}2{\cdot}3{\cdot}4{\cdot}5{\cdot}6) \equiv 1{\cdot}2{\cdot}3{\cdot}4{\cdot}5{\cdot}6 \pmod 7$$

For what values of a are the six numbers $a{\cdot}1$, $a{\cdot}2$, $a{\cdot}3$, $a{\cdot}4$, $a{\cdot}5$ and $a{\cdot}6$ congruent to the six integers 1, 2, 3, 4, 5 and 6 *modulo* 7, in some order (i.e. shuffled about)?

For each such value of a,

$$(a{\cdot}1){\cdot}(a{\cdot}2){\cdot}(a{\cdot}3){\cdot}(a{\cdot}4){\cdot}(a{\cdot}5){\cdot}(a{\cdot}6) \equiv 1{\cdot}2{\cdot}3{\cdot}4{\cdot}5{\cdot}6 \pmod 7$$

and rearranging

$$(a^6){\cdot}(1{\cdot}2{\cdot}3{\cdot}4{\cdot}5{\cdot}6) \equiv 1{\cdot}2{\cdot}3{\cdot}4{\cdot}5{\cdot}6 \pmod 7$$

31 Just shuffling and then power (*modulo* 3, 5, 11)

Fill in the multiplication table *modulo* 3

x	$=$	0	1	2
$0{\cdot}x$	$\equiv$			
$1{\cdot}x$	$\equiv$			
$2{\cdot}x$	$\equiv$			

using only the integers 0, 1 and 2. Which of 0, 1 and 2 is congruent to 2^2 (mod 3)?

Fill in the multiplication table *modulo* 5

x	$=$	0	1	2	3	4
$0{\cdot}x$	$\equiv$					
$1{\cdot}x$	$\equiv$					
$2{\cdot}x$	$\equiv$					
$3{\cdot}x$	$\equiv$					
$4{\cdot}x$	$\equiv$					

using only the integers 0, 1, 2, 3 and 4. Which of 0, 1, 2, 3 and 4 is congruent to 2^4, to 3^4, and to 4^4 (mod 5)?

Fill in the multiplication table *modulo* 11

x	$=$	0	1	2	3	4	5	6	7	8	9	10
$0 \cdot x$	$\equiv$											
$1 \cdot x$	$\equiv$											
$2 \cdot x$	$\equiv$											
$3 \cdot x$	$\equiv$											
$4 \cdot x$	$\equiv$											
$5 \cdot x$	$\equiv$											
$6 \cdot x$	$\equiv$											
$7 \cdot x$	$\equiv$											
$8 \cdot x$	$\equiv$											
$9 \cdot x$	$\equiv$											
$10 \cdot x$	$\equiv$											

using only the integers 0, 1, 2, 3, 4, 5, 6, 7, 8, 9 and 10. Which of these 11 integers is congruent to $2^{10}, 3^{10}, 4^{10}, 5^{10}, 6^{10}, 7^{10}, 8^{10}, 9^{10}$ and to 10^{10}?

If a does not have a factor 3, check that $a \cdot 1$ and $a \cdot 2$ are not congruent either to 0 or to each other *modulo* 3 and therefore one is congruent to 1 and the other is congruent to 2 *modulo* 3.

Deduce that $(a \cdot 1) \cdot (a \cdot 2) \equiv 1 \cdot 2$ (mod 3), just shuffling, and so $a^2 \cdot (1 \cdot 2) \equiv 1 \cdot 2$ (mod 3) giving $a^2 \equiv 1$ (mod 3), to the power of 2.

If a does not have a factor 5, check that $a \cdot 1$, $a \cdot 2$, $a \cdot 3$ and $a \cdot 4$ are not congruent either to 0 or to any one of the others in the list *modulo* 5, so one of these numbers is congruent to each of 1, 2, 3 and 4 *modulo* 5.

Deduce that $(a \cdot 1) \cdot (a \cdot 2) \cdot (a \cdot 3) \cdot (a \cdot 4) \equiv 1 \cdot 2 \cdot 3 \cdot 4$ (mod 5), just shuffling, and so $a^4 \cdot (1 \cdot 2 \cdot 3 \cdot 4) \equiv 1 \cdot 2 \cdot 3 \cdot 4$ (mod 5) giving $a^4 \equiv 1$ (mod 5), to the power of 4.

If a does not have a factor 11, check that $a \cdot 1$, $a \cdot 2$, $a \cdot 3$, . . . , $a \cdot 10$ are not congruent either to 0 or to any one of the others in the list *modulo* 11, so one of these numbers is congruent to each of 1, 2, 3, . . . , 10 *modulo* 11.

Deduce that $(a \cdot 1) \cdot (a \cdot 2) \cdot (a \cdot 3) \cdot \ldots \cdot (a \cdot 10) \equiv 1 \cdot 2 \cdot 3 \cdot \ldots \cdot 10$ (mod 11), just shuffling, and so $a^{10} \cdot (1 \cdot 2 \cdot 3 \cdot \ldots \cdot 10) \equiv 1 \cdot 2 \cdot 3 \cdot \ldots \cdot 10$ (mod 11), giving $a^{10} \equiv 1$ (mod 11), to the power of 10.

32 Factorials

Simplify

1·2·3·4	*modulo* 5
1·2·3·4·5·6	*modulo* 7
1·2·3·4·5·6·7·8·9·10	*modulo* 11

Within each of these products

(i) identify pairs of integers with product ≡ 1 such as 2·4 ≡ 1 (mod 7);
(ii) identify integers whose squares ≡ 1, such as 6·6 ≡ 1 (mod 7).

Look for solutions to x^2 ≡ 1 (mod p) in the multiplication tables for $p = 3$, $p = 5$, $p = 7$ and $p = 11$. What is the general picture? Use $p \mid x^2 − 1 \Rightarrow p \mid x − 1$ or $p \mid x + 1$, for any prime p, to establish your conjecture. Look again at the multiplication tables for $p = 3, 5, 7$ and 11, given in the comments on **30.** and **31. Just shuffling and then** ^power^.

Is there a one in every row? Might there be two ones in a row? Is there a one in every column? Might there be two ones in a column? Where is the one in the last row? Apart from the one in the first row and the last row, are there any more ones on the main diagonal from top left to bottom right?

If ab ≡ 1 (mod p), is ba ≡ 1 (mod p)? When p is an odd prime number, is there an odd or an even number of numbers in the list 2, 3, 4, . . . , $p − 2$? Simplify their product *modulo p*. Can you now get $(p − 1)!$ ≡ $p − 1$ ≡ −1 (mod p)?

33 Square roots of -1, prime modulus

Which of 0, 1 and 2 are solutions of $x^2 + 1$ ≡ 0 (mod 3)? Which of 0, 1, 2, 3 and 4 are solutions of $x^2 + 1$ ≡ 0 (mod 5)?

Look down the main diagonal (top left to bottom right) of the multiplication tables for prime modulus in the comments on **30.** and **31. Just shuffling and then** ^power^ and search for the entry $p − 1$ (mod p). If you find such an entry, use it to construct a solution to $x^2 + 1$ ≡ 0 (mod p).

List the prime moduli that you find for which the square of an integer ≡ −1 (mod p). Also list the prime moduli for which no square of an integer ≡ −1 (mod p).

When there is a solution to $x^2 + 1$ ≡ 0 (mod p) is it ≡ ±1·2·3· . . . ·$\frac{1}{2}(p − 1)$ (mod p)?

34 How many square roots of -1?

Are there any solutions of $x^2 + 1 \equiv 0$ (mod 15)? Try proving that there are no solutions of $x^2 + 1 \equiv 0$ (mod $3n$), and then that there are no solutions of $x^2 + 1 \equiv 0$ (mod pn) where p is a prime number $\equiv 3$ (mod 4).

Look at the two solutions you have found to $x^2 + 1 \equiv 0$ (mod 5). What is the sum of these two solutions? Find two solutions of $x^2 + 1 \equiv 0$ (mod 13). Once you have found one, can you predict the other? You know that $4^2 + 1 \equiv 0$ (mod 17). What is another solution of $x^2 + 1 \equiv 0$ (mod 17)? You know that $12^2 + 1 \equiv 0$ (mod 29). What is another solution of $x^2 + 1 \equiv 0$ (mod 29)?

If n is an odd number and $x^2 + 1 \equiv 0$ (mod n) has a solution, must it have at least two solutions?

$x = 1$ is a solution of $x^2 + 1 \equiv 0$ (mod 2) and $x = 2$ is a solution of $x^2 + 1 \equiv 0$ (mod 5). Use **21. Chinese remainders** to find a solution of $x^2 + 1 \equiv 0$ (mod 10). Find a second solution of $x^2 + 1 \equiv 0$ (mod 10).

In each case where you have exactly two solutions to $x^2 + 1 \equiv 0$ (mod n), does the congruence $x^2 + 1 \equiv (x - \text{first solution})(x - \text{second solution})$ (mod n) hold for all integers x?

Which of 8, 18, 47 and 57 are solutions of $x^2 + 1 \equiv 0$ (mod 65)? Is $x^2 + 1 \equiv (x - 8)(x - 57)$ (mod 65)? Try putting $x = 18$ in both sides of this congruence. Does it still hold, even though neither factor on the right $\equiv 0$ (mod 65)? How do you explain this possibility?

35 Sums of squares

On square dotted paper draw some squares with lattice points as their corners. What are the areas of your squares if we assume the shortest distance between any two lattice points is 1?

Do any of your squares have area 4? Is it possible to draw such a square?
Do any of your squares have area 5? Is it possible to draw such a square?
Do any of your squares have area 6? Is it possible to draw such a square?
The points on the dotted paper can be given integer coordinates.

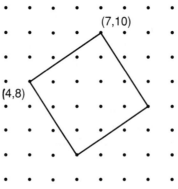

Figure 4

On the dotted paper the two points, (4,8) and (7,10), were chosen as one side of a square. What is the area of this square? Can you relate the area to the coordinates of the two points (4,8) and (7,10)? On one of your own squares write down the coordinates of two corners of one side of the square. See if you can obtain the area of the square from the coordinates of these two corners.

Can you draw a square on the dotted paper with its corners at spots, with area $2^2 + 2^2$?

You will already have found that you can easily draw squares with areas 4 and 5 but that there does not appear to be one of area 6.

Which numbers can be the areas of squares with corners on the lattice?

Can you prove that no square on this lattice has area 6?

To investigate this further we shall look at some sums of squares. First fill in the table giving all possible sums of squares for numbers less than or equal to 99. (There is no need to fill in the table above the diagonal as the results are symmetrical.)

+	0	1	4	9	16	25	36	49	64
0		*	*	*	*	*	*	*	*
1			*	*	*	*	*	*	*
4				*	*	*	*	*	*
9					*	*	*	*	*
16						*	*	*	*
25							*	*	*
36								*	*
49									*
64									
81									

Which numbers less than 100 are sums of squares? Mark them on the display below.

To help study sums of squares we first need to see what sorts of numbers are squares. Using a different colour circle all the square numbers between 0 and 99 in the display. Which columns do they come in? Try adding together any two numbers from the columns with squares. Can you ever get a number from the last column? Look back to see what happened in **16. No shuffling**. Labelling the columns as $4k$, $4k + 1$, $4k + 2$ and $4k + 3$ will help make this clearer.

0	1	2	3
4	5	6	7
8	9	10	11
12	13	14	15
16	17	18	19
20	21	22	23
24	25	26	27
28	29	30	31
32	33	34	35
36	37	38	39
40	41	42	43
44	45	46	47
48	49	50	51
52	53	54	55
56	57	58	59
60	61	62	63
64	65	66	67
68	69	70	71
72	73	74	75
76	77	78	79
80	81	82	83
84	85	86	87
88	89	90	91
92	93	94	95
96	97	98	99

Which columns do not have any sums of squares? Can you give a reason for this?

Odd primes are either of the form $4k + 1$ or $4k + 3$. Make a conjecture about the sort of primes that may be sums of squares.

36 Sums of squares in two ways

A toy manufacturer wants to market a box containing at least 150 toy soldiers. The number of soldiers must be a sum of two squares so that all the soldiers in one box can be used to form two different sized square formations. Below is an example of two different square formations with 20 soldiers.

Figure 5

However, this is not good enough for the manufacturer as it is the only way to make two different squares with 20 soldiers and the manufacturer would like there to be at least two different formations of two squares but using all the soldiers each time. What is the smallest number of soldiers that the manufacturer needs to put into each box?

We shall first investigate which numbers can be written as the sums of squares in two ways. Looking back at the table that you completed in **35. Sums of Squares** find all the numbers that appear twice.

To answer the manufacturer's question we need to discover why some numbers can be written as the sum of two squares in two different ways. Check that

$$(a^2 + b^2)(c^2 + d^2) = (ac + bd)^2 + (ad - bc)^2$$

For what integers a, b is $a^2 + b^2 = 5$? For what c, d is $c^2 + d^2 = 13$? Now use this equation to find 65 as a sum of squares and then use the freedom you had to choose a and b to find 65 as a different sum of squares.

Now try to do the same for 85 and 145.

Look at the prime factors of 65, 85 and 145. Are any of them of the form $4k + 3$? Which prime factors do you think give sums of squares?

Can you now answer the toy manufacturer's question?

37 Pythagorean triples

Because $3^2 + 4^2 = 5^2$, and 3, 4 and 5 are positive integers, (3 4 5) is called a Pythagorean triple after the theorem of Pythagoras: when a right-angled triangle has a hypotenuse of length c and perpendicular sides of length a and b then $a^2 + b^2 = c^2$. Can you use the triple (3 4 5) to make some more Pythagorean triples? Apart from (3 4 5), might there be another Pythagorean triple (a b c) made of consecutive integers?

If a and b have a common factor d, and (a b c) is a Pythagorean triple, must c also have a factor d? If a and c have a common factor d, and (a b c) is a Pythagorean triple, must b also have a factor d?

When (a b c) is a Pythagorean triple and the numbers a, b and c have no common factor greater than 1, then (a b c) is called a *primitive* Pythagorean triple. Here are some primitive Pythagorean triples (a b c) with $a < b < c$: (3 4 5); (5 12 13); (8 15 17); (7 24 25); (20 21 29); (12 35 37); (9 40 41); (28 45 53); (11 60 61); (16 63 65); (33 56 65). Check through this list. If $c = b + 1$, what can you say about a? Can you prove it? ($b = a + 1$ is trickier.)

Look for multiples of 3 in the primitive Pythagorean triples above. Is there any kind of pattern? Complete the following table to show the possibilities for $a^2 + b^2$ (mod 3):

mod 3			b	0	1	2
			b^2	0	1	1
a	a^2	$+$				
0	0					
1	1					
2	1					

What are the possibilities for c^2 (mod 3)? Now use the equation $a^2 + b^2 = c^2$ to see why either a or $b \equiv 0$ (mod 3).

Look for multiples of 5 in the primitive Pythagorean triples above. Is there any kind of pattern? Examine $a^2 + b^2$ (mod 5) with a table, as you did for modulo 3, and c^2 (mod 5) to see if you can establish the pattern you found.

Look for odd and even numbers in the primitive Pythagorean triples above. Which of a, b and c may be which? Look for multiples of 4 in the list of triples above. Is there any kind of pattern? Examine $a^2 + b^2$ (mod 8) and c^2 (mod 8).

Make a table to show the possibilities for $a^2 + b^2$ (mod 8). Look first at what would happen if both a and b were odd numbers. Can you get $a^2 + b^2 = c^2$? Then look at the remaining possibilities to see why either a or b must have a factor 4. What can you say about c in a primitive Pythagorean triple?

Find two positive integers p and q such that $p^2 + q^2 = 5$. Work out $p^2 - q^2$ and $2pq$. Have you resurrected the triple (3 4 5)?

For each of the primitive Pythagorean triples given above, can you find p and q such that $p^2 + q^2 = c$? If you can, work out $2pq$ and $p^2 - q^2$. Are these the other numbers in the triple? For any integers p and q, must $(p^2 - q^2 \; 2pq \; p^2 + q^2)$ be a Pythagorean triple? Use this possibility to search for a Pythagorean triple with $c = 85$. Do you think all primitive Pythagorean triples may be expressed in this form? Try working with the equation $b^2 = (c - a)(c + a)$ and the supposition that a and c are both odd. If you can find a reason why $\frac{1}{2}(c - a)$ and $\frac{1}{2}(c + a)$ should both be squares, then you can take one as p^2 and the other as q^2.

If c is odd and $c = p^2 + q^2$, what can you say about c (mod 4)?

So what are the possible primitive Pythagorean triples *modulo* 4?

PART III

PART III

38 Squares and non-squares

$1^2 \equiv 6^2 \equiv 1 \pmod 7$, $2^2 \equiv 5^2 \equiv 4 \pmod 7$, $3^2 \equiv 4^2 \equiv 2 \pmod 7$. So we will call 1, 2 and 4 *squares modulo* 7, and 3, 5 and 6, *non-squares*.

	squares	non-squares
modulo 3	1	2
modulo 5	1, 4	2, 3
modulo 7	1, 2, 4	3, 5, 6
modulo 11	1, 3, 4, 5, 9	2, 6, 7, 8, 10

Justify each of the entries in the *squares* column.

Make multiplication tables *modulo* 3, 5, 7 and 11 separating the *squares* from the *non-squares*.

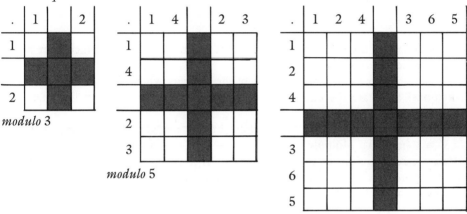

modulo 3

modulo 5

modulo 7

modulo 11

From these tables, what would you predict about

   (i)   the product of two *squares*,
  (ii)   the product of a square and a *non-square*,
 (iii)   the product of two *non-squares*?

Try to justify the first of these claims, and if you are successful, try the other two.

39 Powers of squares and non-squares

Work out all the squares *modulo* 5, and then decide for which x are $x^2 \equiv 1$ and for which x are $x^2 \equiv -1$ (mod 5).

    Work out all the cubes *modulo* 7, and then decide for which x are $x^3 \equiv 1$ and for which x are $x^3 \equiv -1$ (mod 7).

    Work out all the fifth powers *modulo* 11, and then decide for which x are $x^5 \equiv 1$ and for which x are $x^5 \equiv -1$ (mod 11).

    Look for a pattern. Can you classify the integers which satisfy each of the two equations? How are the equations related to the modulus? Predict the analogous results *modulo* 13.

    Multiply out $(x - 1)(x - 4)$ and then simplify *modulo* 5. Deduce two solutions of $x^2 \equiv 1$ (mod 5). Multiply out $(x - 1)(x - 2)(x - 4)$ and then simplify *modulo* 7. Deduce three solutions of $x^3 \equiv 1$ (mod 7). Why must the *squares modulo* 11 provide five solutions of $x^5 \equiv 1$ (mod 11)? What theorem can you claim to help you?

    Don't multiply out $(x - 1)(x - 3)(x - 4)(x - 5)(x - 9)$ or try to simplify this expression. Give a reason why there must be integers a, b, c, d and e such that

$$\begin{aligned}
x^5 - 1 &= (x - 1)(x - 3)(x - 4)(x - 5)(x - 9) \\
&\quad + a(x - 1)(x - 3)(x - 4)(x - 5) \\
&\quad + b(x - 1)(x - 3)(x - 4) \\
&\quad + c(x - 1)(x - 3) \\
&\quad + d(x - 1) \\
&\quad + e
\end{aligned}$$

for all values of x (without regard to modular arithmetic). Now put $x = 1$ to find e. The numbers 3, 4, 5 and 9 are squares *modulo* 11. Put $x = 3$ to find the value of d (mod 11). Put $x = 4$ to find the value of c (mod 11). Put $x = 5$ to find the value of b (mod 11). Put $x = 9$ to find the value of a (mod 11). So what would you have got if you had multiplied out $(x - 1)(x - 3)(x - 4)(x - 5)(x - 9)$ and simplified *modulo* 11? Give a reason why a non-square cannot satisfy $x^5 \equiv 1$ (mod 11). Deduce five solutions of $x^5 \equiv -1$ (mod 11).

40 The frequency of factors

For each of the numbers

$$x = 1 \ 2 \ 3 \ 4 \ 5 \ 6 \ 7 \ 8 \ 9 \ 10 \ 11 \ 12$$

write down hcf(12, x):

	number of xs
when hcf(12, x) = 1; x =	4 = $\phi(12)$
when hcf(12, x) = 2; x =	2
when hcf(12, x) = 3; x =	2
when hcf(12, x) = 4; x =	2
when hcf(12, x) = 6; x =	1
when hcf(12, x) = 12; x =	1

What possible values may hcf(12, x) have? How do these numbers relate to 12?

Add up the numbers in the right hand column. What is the result? Had you expected this?

Euler's ϕ function gives us a way of counting how many integers x there are between 1 and 12, which have hcf(12, x) = 1. Can we use Euler's function to give us a way of counting how many integers x there are between 1 and 12, which have hcf(12, x) = 2? If hcf(12, x) = 2, then x has a factor 2, so what is the value of hcf(6, $x/2$)? This allows us to count the xs such that hcf(12, x) = 2 and get exactly $\phi(6)$ of them.

Now use the ϕ function to count how many integers x there are between 1 and 12 such that hcf(12, x) = 3. Carry on until you have a nice justification of the equation

$$\phi(1) + \phi(2) + \phi(3) + \phi(4) + \phi(6) + \phi(12) = 12$$

Divide each of the numbers 1, 2, 3, 4, 6, 12 into 12. What do you get?

41 Multiplication like addition

addition, *modulo* 4				
+	0	1	2	3
0				
1				
2				
3				

multiplication, *modulo* 5				
.	1	2	4	3
1				
2				
4				
3				

Make entries in the first table with the integers 0, 1, 2 and 3 in such a way that the sums are congruent to the entries *modulo* 4. Make entries in the second table with the integers 1, 2, 3 and 4 in such a way that the products are congruent to the entries *modulo* 5. If the numbers in the two tables are coloured

left		right
0	*yellow*	1
1	*red*	2
2	*green*	4
3	*blue*	3

are the two colour schemes on the tables identical?

Check that in the addition table, the four different entries are congruent to 1, $1 + 1$, $1 + 1 + 1$ and $1 + 1 + 1 + 1$ *modulo* 4. Also check that in the multiplication table the four entries are congruent to 2, 2·2, 2·2·2 and 2·2·2·2 *modulo* 5.

Of the four integers, 0, 1, 2 and 3, which can be used by itself to reach all the others in the additive group *modulo* 4 by repeated addition of the starting number? Those numbers with which you are successful are called the *generators* of this group. In the case of each of the generators, x, what is hcf(4, x)?

Of the four integers 1, 2, 4 and 3, which can be used by itself to reach all the others by repeated multiplication of the starting number *modulo* 5? Those numbers with which you are successful are called the *generators* of this group. Are the generators of the two groups colour matched?

Set up an addition table for the six integers 0, 1, 2, 3, 4 and 5 *modulo* 6, and a multiplication table for the six integers 1, 2, 3, 4, 5 and 6 *modulo* 7. Find a generator in each of these groups. Use these generators to find a rearrangement of the numbers 1, 2, 3, 4, 5 and 6 (mod 7) which provides the possibility of a colour matching of the addition table *modulo* 6 and the multiplication table *modulo* 7. If x is a generator of the additive group *modulo* 6, what is the value of hcf(6, x)?

Set up an addition table for the ten integers 0, 1, 2, . . . , 9 *modulo* 10, as you did in **12. Spot check**, and a multiplication table for the 10 integers 1, 2, 3, . . . , 10 *modulo* 11, as you did in **31. Just shuffling and then** [power]. Find a generator for each group, and then show that a perfect colour matching is possible.

Identify the four generators in the additive group *modulo* 10, and use them to find the four generators of the multiplicative group *modulo* 11.

42 Powers to a prime modulus

modulo 5

x	x^2	x^3	x^4	x^5
0	0	0	0	0
1	1	1	1	1
2	4	3	1	2
3	4	2	1	3
4	1	4	1	4

The last two columns exhibit Fermat's theorem in the form $x^4 \equiv 1 \pmod{5}$ for x not $\equiv 0 \pmod{5}$, and in the form $x^5 \equiv x \pmod{5}$ for all x. In this context (that of multiplication *modulo* 5) the *order* of an integer is the name given to the smallest positive index n such that $x^n \equiv 1 \pmod{5}$.

What is the order of 2, of 3, of 4 and of 1? How many of 1, 2, 3 and 4 are of order 1, of order 2 and of order 4? Compare these numbers with $\phi(1)$, $\phi(2)$ and $\phi(4)$ to illustrate the equation $\phi(1) + \phi(2) + \phi(4) = 4$, as established in **section 40**.

Make a table of powers *modulo* 7.

x	x^2	x^3	x^4	x^5	x^6	x^7
0						
1						
2						
3						
4						
5						
6						

Do the last two columns again exhibit the two forms of Fermat's theorem, as given in the comment on **31. Just shuffling and then** power? What is the order of 3, of 5, of 2, of 4, of 6 and of 1 under multiplication *modulo* 7? What must the order of a generator be? (The term *generator* is defined in the comment on **section 41**.)

How many of 1, 2, 3, 4, 5 and 6 are of order 1, of order 2, of order 3 and of order 6? Compare these numbers with $\phi(1)$, $\phi(2)$, $\phi(3)$ and $\phi(6)$, to illustrate the equation $\phi(1) + \phi(2) + \phi(3) + \phi(6) = 6$ from **section 40**.

Make a table of powers *modulo* 11.

x	x^2	x^3	x^4	x^5	x^6	x^7	x^8	x^9	x^{10}	x^{11}
0										
1										
2										
3										
4										
5										
6										
7										
8										
9										
10										

Do the last two columns exhibit the two forms of Fermat's theorem? What is the order of 2, of 6, of 7 and of 8 under multiplication *modulo* 11? What is the order of 3, of 4, of 5 and of 9? What is the order of 10, and what is the order of 1?

How many of 1, 2, 3, 4, 5, 6, 7, 8, 9 and 10 are of order 1, of order 2, of order 5 and of order 10? Compare these numbers with $\phi(1)$, $\phi(2)$, $\phi(5)$ and $\phi(10)$ to illustrate the equation $\phi(1) + \phi(2) + \phi(5) + \phi(10) = 10$ from **section 40**.

Can you predict the order of a generator for multiplication *modulo p*? Make a list of all the orders which occur for each prime modulus. How do these orders relate to the order of a generator?

43 Zero products

Make multiplication tables *modulo* 4, *modulo* 6, *modulo* 8 and *modulo* 12.

Where do the zeros come in these tables? Without making the multiplication table, predict a product of zero in the multiplication table for *modulo* 39.

Make a list of the integers a with $0 < a < 12$ such that the list $a \cdot 1, a \cdot 2, a \cdot 3, \ldots,$ $a \cdot 11$ contains an integer $\equiv 0 \pmod{12}$. If $\mathrm{hcf}(12, a) = 1$, is there a zero product in the list? If $\mathrm{hcf}(12, a) = d > 1$, can you predict, and then find, a zero product in the list?

44 Non-zero products

Make multiplication tables for {1, 3} *modulo* 4,
{1, 5} *modulo* 6,
{1, 3, 5, 7} *modulo* 8,
{1, 2, 4, 5, 7, 8} *modulo* 9.

For which a taken from the set {1, 2, 3, 4, 5, 6, 7, 8} does the set {$a \cdot 1$, $a \cdot 2$, $a \cdot 3$, $a \cdot 4$, $a \cdot 5$, $a \cdot 6$, $a \cdot 7$, $a \cdot 8$} contain no integer $\equiv 0$ (mod 9)?

If a is taken from the set {1, 2, 4, 5, 7, 8}, does the list $a \cdot 1$, $a \cdot 2$, $a \cdot 4$, $a \cdot 5$, $a \cdot 7$, $a \cdot 8$

(i)   contain an integer $\equiv 1$ (mod 9);
(ii) contain integers congruent to each of 1, 2, 4, 5, 7 and 8 *modulo* 9?

For which a taken from the set {1, 2, . . . , 9} does the set {$a \cdot 1$, $a \cdot 2$, . . . , $a \cdot 9$} contain no integer $\equiv 0$ (mod 10)? If a is taken from the set {1, 3, 7, 9}, does the list $a \cdot 1$, $a \cdot 3$, $a \cdot 7$, $a \cdot 9$

(i)   contain an integer $\equiv 1$ (mod 10);
(ii) contain integers congruent to each of 1, 3, 7 and 9 *modulo* 10?

Pose and answer analogous questions in arithmetic *modulo* 12.

45 Decimals to the death

The fraction 1365/4515 has *numerator* = 1365 and *denominator* = 4515:

$$\frac{1365}{4515} = \frac{455}{1505} = \frac{91}{301} = \frac{13}{43}$$

so the fraction, in lowest terms, is 13/43.

Terminators

What happens every time if you write the following fractions as decimals: 1/2, 3/5, 43/20, 7/25, 137/50, 1003/500, 1003/400, 51/32?

What do you notice about the denominators when the following decimals are expressed as fractions in their lowest terms: 3.5, 4.7, 0.12, 2.45, 0.375, 1.054, 30.3125, 0.16?

Conjecture what fractions are equal to terminating decimals and vice versa. Focus on the denominators, and their prime factors. Attempt to justify your conjecture; that is, prove your claim.

Cell block 1

What happens every time if you write the following fractions as decimals: 1/3, 2/3, 1/6, 5/6, 2/9, 8/9, 17/12, 17/15, 7/18, 29/24?

What do you notice about the denominators when the following decimals are expressed as fractions in their lowest terms: $1.\dot{3}$ (which means $1.333\ldots$, with the 3 recurring), $0.0\dot{3}$, $0.00\dot{3}$, $1.1\dot{4}$, $2.2\dot{3}$, $0.\dot{1}$, $0.\dot{9}$, $0.1234\dot{5}$? (If the decimal you are seeking to interpret is x, write down the decimal for $10x$, and then calculate $10x - x$ which you should find, in these cases, always terminates.)

Conjecture what fractions are equal to decimals with a single recurring digit ($\neq 0$ or 9). Again, focus on the denominators and their prime factors. Attempt to justify your conjecture. What fractions are equal to decimals with a single recurring nine? Try $0.4\dot{9}$.

Cell block 2

What happens every time if you write the following fractions as decimals: $2/11$, $3/22$, $4/33$, $5/44$, $6/55$, $7/66$, $9/88$, $10/99$, $101/176$?

What do you notice about the denominators when the following decimals are expressed as fractions in their lowest terms: $0.\dot{0}\dot{9}$ (which means $0.09090909\ldots$, with the 09 recurring), $1.\dot{1}\dot{2}$, $1.8\dot{1}\dot{2}$, $1.87\dot{1}\dot{2}$, $1.0\dot{2}\dot{7}$, $2.12\dot{1}\dot{6}$? (If the decimal you are seeking to interpret is x, write down the decimal for $100x$, and then calculate $100x - x$.)

Conjecture what fractions are equal to decimals with a pair of recurring digits. Attempt to justify your conjecture.

Cell block 6

What happens every time if you write the following fractions as decimals: $1/7$, $2/7$, $3/7$, $1/13$, $2/13$, $3/14$, $5/21$, $3/26$, $13/28$, $8/77$?

Factorise $999,999$ into a product of prime numbers.

What do you notice about the denominators when the following decimals are expressed as fractions in their lowest terms: $0.\dot{7}1428\dot{5}$, $1.\dot{2}3076\dot{9}$, $0.\dot{3}5714\dot{2}\dot{8}$, $0.\dot{3}8095\dot{2}$, $0.1\dot{9}2307\dot{6}$, $0.\dot{1}2820\dot{5}$? (If the decimal you are seeking to interpret is x, write down the decimal for $1000000x$, and then calculate $1000000x - x$.)

Conjecture what fractions are equal to decimals with a block of six recurring digits. Attempt to justify your conjecture. Can you say how to find a fraction equal to a recurring decimal with a recurring block of length n?

Optional: Here is a puzzle in which each letter stands for a different digit. Find the digits.

$$\frac{T}{MY} = .REPEATREPEATREPEAT\ldots$$

46 Recurring decimals

The patterns you found in **45. Decimals to the death** show that it is the denominator of a fraction that determines the length of the recurring block in the equivalent decimal expression, and that the key to working out the length of the recurring block comes from the calculation that if a, b and c are digits from the set $\{0, 1, 2, \ldots, 9\}$ then

$$\frac{a}{9} = 0.\dot{a}, \quad \frac{ab}{99} = 0.\dot{a}\dot{b}, \quad \frac{abc}{999} = 0.\dot{a}b\dot{c}$$

and so on.

This lets us get a wider picture if only we know the factors of the integers with nines as their only digits. Here are the first few:

$$9 = 3 \cdot 3 \qquad\qquad 99{,}999 = 3 \cdot 3 \cdot 41 \cdot 271$$
$$99 = 3 \cdot 3 \cdot 11 \qquad\qquad 999{,}999 = 3 \cdot 3 \cdot 3 \cdot 7 \cdot 11 \cdot 13 \cdot 37$$
$$999 = 3 \cdot 3 \cdot 3 \cdot 37 \qquad\qquad 9{,}999{,}999 = 3 \cdot 3 \cdot 239 \cdot 4649$$
$$9{,}999 = 3 \cdot 3 \cdot 11 \cdot 101 \qquad\qquad 99{,}999{,}999 = 3 \cdot 3 \cdot 11 \cdot 101 \cdot 73 \cdot 137$$

Express each of these fractions as a recurring decimal:

$$\frac{12345}{99999}, \ \frac{1234}{99999}, \ \frac{123}{99999}, \ \frac{521}{11111}, \ \frac{1}{9 \cdot 41}, \ \frac{1}{410}, \ \frac{1}{82}, \ \frac{1}{164}, \ \frac{1}{205}, \ \frac{1}{1025}$$

Use a calculator as little as possible for the calculations or you will miss the point.

Use a calculator the find a multiple of the prime number 9091 each of whose digits are nines. (From the list above it must have at least nine digits.) Express 1/9091 as a recurring decimal. After doing this write down some other fractions whose recurring decimal form has the same length of recurring block as 1/9091.

17 is a factor of 9,999,999,999,999,999. If you were to express 1/17 as a recurring decimal how long might the recurring block be? Can you think of a reason (not just a calculation) why 17 should be a factor of 9,999,999,999,999,999? Fermat's theorem should help.

If an integer n does not have a factor 2 or 5, we can find the recurring decimal for m/n if only we can find an equivalent fraction with all the digits in the denominator nines. If $1/n$ is expressed as a decimal find out (in as many cases as you can) the length of the recurring block for $n = 1$ to 50. Do you get the same length of recurring block for $2/n$? In which cases do they differ? What about $3/n$? Just check the feasible cases up to $n = 50$.

Fill in the values in the following table:

$R(n) = $ length of the recurring block in the decimal expression for $1/n$.

$R(3) =$	,	$= \phi(3)$	$R(6) =$	,	$= \phi(3)$	$R(7) =$	,	$= \phi(7)$
$R(9) =$	,	$= \phi(9)$	$R(11) =$	,	$= \phi(11)$	$R(12) =$	,	$= \phi(3)$
$R(13) =$	,	$= \phi(13)$	$R(14) =$	,	$= \phi(7)$	$R(15) =$	,	$= \phi(3)$
$R(17) =$	,	$= \phi(17)$	$R(18) =$	,	$= \phi(9)$	$R(21) =$	,	$= \phi(21)$
$R(22) =$	,	$= \phi(11)$	$R(24) =$	,	$= \phi(3)$	$R(26) =$	,	$= \phi(13)$
$R(27) =$	,	$= \phi(27)$	$R(28) =$	,	$= \phi(7)$	$R(30) =$	,	$= \phi(3)$

Can you see a connection between the n of $R(n)$ above and the selected m for $\phi(m)$ next to it? Conjecture a reason for this choice.

Can you see any relation between the pairs of numbers which you have computed in the table above?

What do the digits of the number $10^n - 1$ look like? Express the statement "m is a factor of $10^n - 1$" using a congruence.

If p is a prime number (different from 2 and 5) is there always a value of n for which $10^n \equiv 1 \pmod{p}$? If $p \mid 10^n - 1$, must there be an integer k such that $1/p = k/(10^n - 1)$? What does this tell you about the length of the recurring block in the decimal expression for $1/p$?

In particular cases (e.g. $p = 13$) might there ever be a smaller n than the one guaranteed by Fermat's theorem such that $10^n \equiv 1 \pmod{p}$?

$$\frac{5}{6} = \frac{5}{2 \cdot 3} = \frac{5 \cdot 5}{2 \cdot 5 \cdot 3} = \left(\frac{1}{10}\right)\left(\frac{25}{3}\right) = \left(\frac{1}{10}\right)\left(8 + \frac{1}{3}\right) = \left(\frac{1}{10}\right)\left(8 + \frac{3}{9}\right) = \left(\frac{1}{10}\right)(8.\dot{3}) = 0.8\dot{3}$$

With the calculation above as a pattern, first try to separate powers of 2 and 5 in the denominator as a stage in working out $7/550$ as a recurring decimal. At one stage along the way you should get

$$\left(\frac{1}{100}\right)\left(1 + \frac{3}{11}\right) = \left(\frac{1}{100}\right)\left(1 + \frac{27}{99}\right)$$

If you have been given a fraction N/M, can you always find an equivalent fraction $L/10^n K = N/M$, where K is an integer without a factor 2 or a factor 5? Can you be sure that some multiple of K is an integer with all its digits nines?

The Fermat–Euler theorem in **44. Non-zero products** will ensure that $10^{\phi(K)} \equiv 1 \pmod{K}$, when $\mathrm{hcf}(10, K) = 1$, to give a recurring decimal for L/K.

Try to give a succinct summary of the work in this section. What integers are factors of at least one number in the sequence 9, 99, 999, 9999, 99999, 999999, . . . ?

47 Can you reveal all and keep it secret?

One of the difficulties of keeping a code secret is that the sender and receiver have to agree beforehand on what code they will use. Ideally we would like people to be able to send each other messages without ever having to meet to arrange a code. Suppose we could find a system where everybody announced publicly how to encode messages to them but how to decode still remained secret. Seems unlikely? The following application of number theory shows that just such a system is possible.

For ease of calculation we shall restrict ourselves to being able to send messages which consist only of the numbers 0, . . . , 9. This time, rather than using multiplication and addition to encode as in **24. Coding and decoding** we will use powers.

Complete the following table:

x	0	1	2	3	4	5	6	7	8	9
x^3 (mod 11)	0	1			9				6	3
$(x^3)^7$ (mod 11)	0	1			4				8	9

What do you notice about the row x^3 *modulo* 11? What do you notice about the row $(x^3)^7$ *modulo* 11?

Suppose the message we want to send consists of the digits 5 0 3 1. Using the table, encode the message by cubing each digit *modulo* 11. In order to decode the message we take the seventh power of each encoded number. (Of course as our table is so small it is just as easy to decode by working back to see which number cubed gives the coded number.) Which numbers can we use as the encoding powers? Will any numbers do?

Complete the following table:

x	0	1	2	3	4	5	6	7	8	9
x^2 (mod 11)	0	1			5				9	4

Is 2 a suitable encoding power *modulo* 11? For *modulo* 11 decide which numbers are suitable encoding powers (i.e. those that give all answers different). Are the numbers you have found coprime to $\phi(11)$? Are they all the numbers less than $\phi(11)$ and coprime to $\phi(11)$?

Now let us consider the decoding power. Why is $(x^3)^7 \equiv x$ *modulo* 11? You may need to recall Fermat's theorem from **31. Just shuffling and then** ^{power}. If we use 5 as the encoding power and work *modulo* 13, what will be a suitable decoding power?

If we want to make this system secure we need to stop anyone who knows the modulus m and the encoder a from determining the decoder b. An efficient decoder b satisfies the equation $ab \equiv 1$ (mod $\phi(m)$). In the comments on **section 44** we saw that if a is coprime to $\phi(m)$ then the elements $a \cdot b_1, a \cdot b_2, \ldots, a \cdot b_\phi(m)$ are the elements 1, 2, $\ldots$, $\phi(m)$ in some order. Hence one of the bs, say b_j, satisfies $a \cdot b_j \equiv 1$ (mod $\phi(m)$).

If $m = 35$ and $a = 7$ then by using the prime factorisation of 35 and the fact that ϕ is multiplicative from **22. Systematic catching practice** find b. The number b can be found by trying the possibilities until a solution of $7b \equiv 1$ (mod 24) appears.

Let $a = 3$ and $m = 2027651281$ and try to determine a value for b. Here the difficult part is to obtain $\phi(m)$. If you know that

$$2027651281 = 44021 \cdot 46061$$

can you now determine $\phi(m)$ and hence b?

Computers can create 100 digit primes very quickly but cannot *at present* factorise the product of two such primes, a 200 digit number, in less than 3.8×10^9

years. Even a 100 digit number can take 74 years to factorise! This leads us to an unbreakable code with both the modulus *m* and the encoder *a* being known to everyone and only the person for whom the message is intended knowing the decoder *b*.

This is the essence of the Public Key System known as the RSA system after its inventors R Rivest, A Shamir and L Adleman.

Note

This code can still be cracked by studying the frequency of letters in the alphabet and the frequency of letters in the coded message. We would soon find the correspondence just as we did for the code in **24. Coding and decoding**. A solution to this is to take letters in pairs so that, for example, the number 0317 corresponds to the letters *cr*, and then the number 0317 is encoded. Now the code is much harder to crack as the frequency of letters occurring as pairs is not so well defined. Write the message COME HOME as a sequence of four digit numbers. In real codes the blocks of letters taken together are much longer, usually 100 at a time, thereby making it infeasible for a code breaker to try all possible combinations of letters.

48 Primes as squares and non-squares

In **25. Repacking** we saw that a prime number may or may not be congruent to a square depending on the modulus. Using the tables below, find for each modulus which primes are in the same column as a square and which are not. This is similar to what you did in **26. Where have all the squares gone?**.

Modulus 3

0	1	2
3	4	5
6	7	8
9	10	11
12	13	14
15	16	17
18	19	20

Modulus 5

0	1	2	3	4
5	6	7	8	9
10	11	12	13	14
15	16	17	18	19

Modulus 7

0	1	2	3	4	5	6
7	8	9	10	11	12	13
14	15	16	17	18	19	20

Modulus 11

0	1	2	3	4	5	6	7	8	9	10
11	12	13	14	15	16	17	18	19	20	21
22	23	24	25	26	27	28	29	30	31	32

Modulus 13

0	1	2	3	4	5	6	7	8	9	10	11	12
13	14	15	16	17	18	19	20	21	22	23	24	25
26	27	28	29	30	31	32	33	34	35	36	37	38

Now complete the table to help in the investigation to discover which primes are squares for which moduli.

Squares or non-squares

		3	5	7	11	13	17	19
	3	*						
	5		*					
	7			*				
modulus	11				*			
	13					*		
	17	N	N	N	N	Y	*	Y
	19	N	Y	Y	Y	N	Y	*

In the above table we can see which primes are squares for the different moduli. Is it true that if p is a square *modulo q* then q is a square *modulo p*? Circle the pairs where it is not true, that is where the entries are not symmetric about the starred diagonal. Which rows and columns are they in? Label the leading row and column with the numbers 1 and 3. Use 1 if $p \equiv 1 \pmod 4$ and 3 if $p \equiv 3 \pmod 4$.

Make a conjecture about the types of p and q which give symmetric entries and those which do not.

We have seen that 2 is not a square *modulo* 3, 5, 11 and 13 but that it is a square *modulo* 2 and 7. What other moduli have 2 as a square?

We know that if $x^2 \equiv 2 \pmod q$ then q divides $x^2 - 2$ and conversely. So by completing the following table you will find some more primes q for which 2 is a square:

x	2	3	4	5	6	7	8	9	10	11	12	13	14	15	16	17	18	19	20
x^2	4	9																	
$x^2 - 2$	2	7																	
prime factors	2	7																		

Classify these primes *modulo* 8.

49 Counting dots in a rectangle

On dotted paper draw the following rectangle three times but with $p = 13$, $q = 7$; $p = 13$, $q = 17$; $p = 11$, $q = 7$. The corners are $Q(1, 1)$, $R(\frac{1}{2}(p - 1), 1)$, $T(1, \frac{1}{2}(q - 1))$, $S(\frac{1}{2}(p - 1), \frac{1}{2}(q - 1))$) and p and q are odd primes.

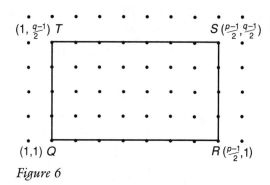

$(1, \frac{q-1}{2})\ T$ $S\ (\frac{p-1}{2}, \frac{q-1}{2})$

$(1,1)\ Q$ $R\ (\frac{p-1}{2},1)$

Figure 6

The lattice points (= integer points) are those points with a pair of integers as co-ordinates. In each case see how many lattice points there are in (and on the sides of) the rectangle. What is the formula for the number of lattice points in terms of p and q? Can you give the conditions under which this formula gives an even number, and when it gives an odd number?

For each of the lattice points (x, y) in the rectangle, calculate $(\frac{1}{2}(p + 1) - x, \frac{1}{2}(q + 1) - y$. Is this point also a lattice point? Where does this point lie?

Can you show that the half turn

$$(x, y) \rightarrow \left(\frac{p + 1}{2} - x, \frac{q + 1}{2} - y \right)$$

maps the rectangle $QRST$ onto itself? Does the half turn always map integer points to integer points? Is it ever possible for the two points to coincide, that is for

$$(x, y) = \left(\frac{p + 1}{2} - x, \frac{q + 1}{2} - y \right)?$$

Now look again at the rectangle with $p = 11$, $q = 7$. How many lattice points were there in the rectangle $QRST$ in this case? Which lattice point in the rectangle is not paired? What can you say about p and q *modulo* 4 if $(\frac{1}{4}(p + 1), \frac{1}{4}(q + 1))$ is a lattice point? Test out your idea by drawing the diagram again but this time with $p = 7$ and $q = 19$.

50 Half-size products

Fill in the columns labelled a, $2a$, $3a$ in the following table *modulo* 7 putting any entries, x, in, which are between $7/2$ and 7 as their numerically least residue *modulo*

7. For example, $3 \cdot 4 \equiv 5 \equiv -2 \pmod 7$, so the number 5 is replaced by -2, writing $x - 7$ for x.

a	a	$2a$	$3a$	n
1				
2				
3				
4				
5				
6				

mod 7

What numbers are in each row? In the last column write down n, the number of negative entries in each row.

Look back at your work on squares *modulo* 7 in **27. Where have all the squares come from?**. What are the squares *modulo* 7 and which values of a in the above table give an even number of negative entries? What are the non-squares *modulo* 7 and which have an odd number of negative entries?

Fill in the columns a, $2a$, $3a$, $4a$, $5a$ in the following table *modulo* 11 putting any entries x between $11/2$ and 11 as $x - 11$, their numerically least residue *modulo* 11. In the last column write down n the number of negative entries in that row:

a	a	$2a$	$3a$	$4a$	$5a$	n
1						
2						
3						
4						
5						
6						
7						
8						
9						
10						

mod 11

Is the relation between squares and non-squares and the number of negative entries the same as that in the table *modulo* 7?

Make a conjecture about the connection between the squares and non-squares *modulo* p and the values of a that lead to rows with even and odd numbers of negative entries.

You may also have noticed that if you disregard the signs the entries in each row of the table are just the numbers 1, 2, 3 in some order for the *modulo* 7 table and 1, 2, 3, 4, 5 in some order for the *modulo* 11 table. What do you think the entries will be for a table of products a, $2a$, $3a$, $4a$, $5a$, $6a$ *modulo* 13 if you disregard the signs? Write out the table and see if your guess is correct.

What do you think the entries will be for a table of products a, $2a$, $3a$, ... , $\frac{1}{2}(p-1)$ a *modulo* p if you disregard the signs? Try to justify this conjecture; that is, that for a prime modulus p, the absolute values of the entries in each row of the table will always be the numbers $1, 2, \ldots, \frac{1}{2}(p-1)$ in some order. If the conjecture is correct then the product of any row of the table will be $\pm(\frac{1}{2}(p-1))!$

From **39. Powers of squares and non-squares** we have seen that $a \cdot \frac{1}{2}(p-1) \equiv +1 \pmod{p}$ if a is a square *modulo* p and $a \cdot \frac{1}{2}(p-1) \equiv -1 \pmod{p}$ if a is a non-square *modulo* p. Can you see how this result ties in with the even and odd number of negative entries? We can combine both these results to show that $a \cdot \frac{1}{2}(p-1) \equiv (-1)^n \pmod{p}$ where n is the number of negative entries in the row.

In the discussion we repeatedly consider the set of integers

$$\{a, 2a, 3a, \ldots, \tfrac{1}{2}(p-1)a\}$$

which we denote by S_a. Sometimes we will need to refer to a set of integers congruent to those in S_a *modulo* p but lying between 0 and p: we will denote this set by $S_a[0, p]$. More often we will refer to a set of integers congruent to those in S_a *modulo* p but lying between $-\frac{p}{2}$ and $\frac{p}{2}$, giving the numerically least residue: we will denote this set by $S_a[-\frac{p}{2}, \frac{p}{2}]$.

In **48. Primes as squares and non-squares** we saw that 2 only seemed to be a square for odd primes $\equiv \pm 1 \pmod 8$. By investigating the number of negative entries (n) in the set

$$S_2[-\tfrac{p}{2}, \tfrac{p}{2}] \equiv \{2, 2.2, 3.2, \ldots, \tfrac{1}{2}(p-1).2\} \pmod{p}$$

you can discover which prime moduli have 2 as a square if you consider the possible moduli *modulo* 8.

51 From dashes to dots

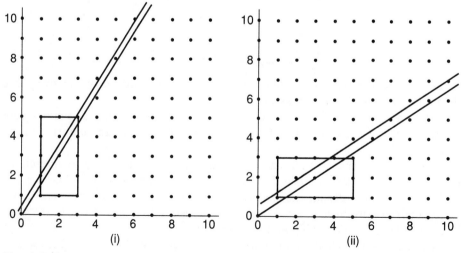

Figure 7(i)–(ii)

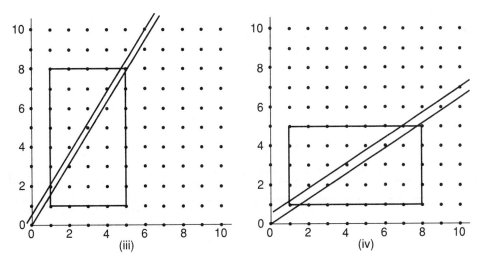

Figure 7(iii)–(iv)

In the figure above we have drawn the lines $y = \frac{q}{p}x$, $y = \frac{q}{p}x + \frac{1}{2}$ for (i) $p = 7$, $q = 11$, (ii) $p = 11$, $q = 7$, (iii) $p = 11$, $q = 17$ and (iv) $p = 17$, $q = 11$. Consider the region A bounded by these lines and $x = 1$ and $x = \frac{1}{2}(p - 1)$. Fill in the following tables:

	x	1	2	$3 = \frac{1}{2}(p - 1)$
	qx	11	22	
	$\frac{qx}{p}$	$\frac{11}{7}$	$\frac{22}{7}$	
$p = 7$	$\frac{qx}{p} - \lfloor \frac{qx}{p} \rfloor$	$\frac{4}{7}$	$\frac{1}{7}$	
$q = 11$	$< \frac{1}{2}, > \frac{1}{2}$	$>$	$<$	
	numerically least residue $qx \bmod p$	-3	1	
	integer between $\frac{qx}{p}$ and $\frac{qx}{p} + \frac{1}{2}$	yes	no	

	x	1	2	3	4	$5 = \frac{1}{2}(p - 1)$
	qx					
	$\frac{qx}{p}$					
$p = 11$	$\frac{qx}{p} - \lfloor \frac{qx}{p} \rfloor$					
$q = 7$	$< \frac{1}{2}, > \frac{1}{2}$					
	numerically least residue $qx \bmod p$					
	integer between $\frac{qx}{p}$ and $\frac{qx}{p} + \frac{1}{2}$					

x	1 2 3 4 5 = $\frac{1}{2}(p-1)$
qx	
qx/p	
$qx/p - \lfloor qx/p \rfloor$	
$< \frac{1}{2}, > \frac{1}{2}$	
numerically least residue qx mod p	
integer between qx/p and $qx/p + \frac{1}{2}$	

$p = 11$
$q = 17$ (shown at left of table)

x	1 2 3 4 5 6 7 8 = $\frac{1}{2}(p-1)$
qx	
qx/p	
$qx/p - \lfloor qx/p \rfloor$	
$< \frac{1}{2}, > \frac{1}{2}$	
numerically least residue qx mod p	
integer between qx/p and $qx/p + \frac{1}{2}$	

$p = 17$
$q = 11$ (shown at left of table)

For a given integer x how many lattice points may there be within the region A? Could there ever be more than one point?

Can you show that there will be such a lattice point if and only if

$$\frac{qx}{p} - \left\lfloor \frac{qx}{p} \right\rfloor > \frac{1}{2}?$$

From **50. Half-size products** we know that in the set $S_q[-\frac{p}{2}, \frac{p}{2}] \equiv \{q, 2q, \ldots,$ $\frac{1}{2}(p-1)q\} \pmod{p}$, the negative entries for $p = 7$ and $q = 11$ are $11 \equiv -3$, and $33 \equiv -2$ *modulo* p, the first and third multiples of q. Does this correspond to the values of x between 1 and $\frac{1}{2}(p-1)$ that give lattice points in the region A?

Try to show that each lattice point in the region A is matched with a negative entry in $S_q[-\frac{p}{2}, \frac{p}{2}]$. In other words show that

$$\frac{qx}{p} - \left\lfloor \frac{qx}{p} \right\rfloor > \frac{1}{2} \Leftrightarrow qx \equiv t \pmod{p} \text{ where } p > t > \frac{p}{2}$$

Are there any lattice points on the lines $y = \frac{q}{p}x$ and $y = \frac{q}{p}x + \frac{1}{2}$ for $1 \le x \le \frac{1}{2}(p-1)$?

52 Quadratic reciprocity

Take the two rectangles $QRST$ which you drew for $(p, q) = (11, 7)$ and $(7, 11)$ in **51. From dashes to dots**. Superimpose origin on origin, x-axis on x-axis and y-axis on *y*-axis. You can do this either by holding two pieces of paper up to the light, or by drawing both rectangles with the same axes. When you have done this, try to identify a mirror line which will give one of the rectangles as a reflection of the other. Does this reflection match lattice points with lattice points throughout the plane? What is the reflection of $y = qx/p$ in this mirror line? Harder. What is the reflection of $y = qx/p + \frac{1}{2}$ in this mirror line?

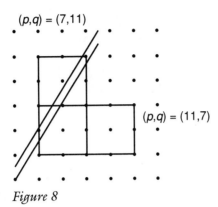

Figure 8

In the rectangle *QRST* for $(p, q) = (11, 7)$, mark the region *A* (as described at the beginning of **section 51**), and in the same rectangle mark the image, *B*, of the region which would have been called *A* for $(p, q) = (7, 11)$ under the reflection which matches the two rectangles. 7 is a non-square *modulo* 11, so *A* contains an odd number (3) of lattice points (as found in **section 51**). 11 is a square *modulo* 7, so *B* contains an even number (2) of lattice points (as found in **section 51**). $(7 \mid 11)$ $\neq (11 \mid 7)$ and $A \cup B$ contains an *odd* number of lattice points.

Now explore the similar situation with $(p, q) = (17, 11)$ and $(11, 17)$ to illustrate $(11 \mid 17) = (17 \mid 11)$ with each a non-square *modulo* the other. Here *A* (with 5) and *B* (with 3) each contain an odd number of lattice points, so together $A \cup B$ contains an even number of lattice points.

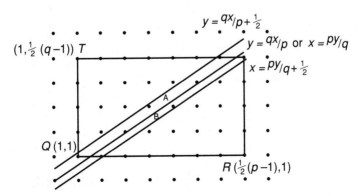

Figure 9

In **49. Counting dots in a rectangle** we considered a half turn which mapped the rectangle $QRST$ onto itself. Does the same half turn map the region $A \cup B$ onto itself? If so, the two regions of $QRST$ outside $A \cup B$ are equally matched, so the numbers of lattice points in $QRST$ and $A \cup B$ are both even or both odd.

If $A \cup B$ contains an odd number of lattice points, what can you say about $(p \mid q)$ and $(q \mid p)$? If $A \cup B$ contains an even number of lattice points, what can you say about $(p \mid q)$ and $(q \mid p)$?

As we found in **49. Counting dots in a rectangle**, the number of lattice points in $QRST$ is $\frac{1}{2}(p-1)\frac{1}{2}(q-1)$, so $QRST$ contains an odd number of lattice points when $p \equiv q \equiv 3 \pmod 4$, and $QRST$ contains an even number of lattice points otherwise.

Can you give a rule for when $(p \mid q) \neq (q \mid p)$ and when $(p \mid q) = (q \mid p)$?

53 Adding squares

Look at the table in the comments on **35. Sums of squares** which exhibits integers expressible in the form $x^2 + y^2$. Look for prime numbers in this table. Classify them *modulo* 4, and try to explain what you have found. How about the non-primes in the table; what are their prime factors like?

$x^2 + 2y^2$

x	x^2	y / y^2 / $2y^2$ +	0 / 0 / 0	1 / 1 / 2	2 / 4 / 8	3 / 9 / 18	4 / 16 / 32	5 / 25 / 50	6 / 36 / 72	7 / 49 / 98	8 / 64 / 128	9 / 81 / 162	10 / 100 / 200
0	0		0	2	8	18	32	50	72	98	128	162	200
1	1		1	3	9	19	33	51	73	99	129	163	201
2	4		4	6	12	22	36	54	76	102	132	166	204
3	9		9	11	17	27	41	59	81	107	137	171	209
4	16		16	18	24	34	48	66	88	114	144	178	216
5	25		25	27	33	43	57	75	97	123	153	187	225
6	36		36	38	44	54	68	86	108	134	164	198	236
7	49		49	51	57	67	81	99	121	147	177	211	249
8	64		64	66	72	82	96	114	136	162	192	226	264
9	81		81	83	89	99	113	131	153	179	209	243	281
10	100		100	102	108	118	132	150	172	198	228	262	300

Look for prime numbers in this table. Classify them *modulo* 8, and try to justify what you have found. How about the non-prime numbers in the table; what are their prime factors like?

$x^2 + 3y^2$

y	0	1	2	3	4	5	6	7	8	9	10
y^2	0	1	4	9	16	25	36	49	64	81	100
$3y^2$	0	3	12	27	48	75	108	147	192	243	300

x	x^2	$+$										
0	0	0	3	12	27	48	75	108	147	192	243	300
1	1	1	4	13	28	49	76	109	148	193	244	301
2	4	4	7	16	31	52	79	112	151	196	247	304
3	9	9	12	21	36	57	84	117	156	201	252	309
4	16	16	19	28	43	64	91	124	163	208	259	316
5	25	25	28	37	52	73	100	133	172	217	268	325
6	36	36	39	48	63	84	111	144	183	228	279	336
7	49	49	52	61	76	97	124	157	196	241	292	349
8	64	64	67	76	91	112	139	172	211	256	307	364
9	81	81	84	93	108	129	156	189	228	273	324	381
10	100	100	103	112	127	148	175	208	247	292	343	400

Look for prime numbers in this table. Classify them *modulo* 6, and try to decide what you would need to know to justify the pattern which you have found. How about the non-prime numbers?

Comments and solutions

PART I

PART I

1c The magic of nines – comments

In the nine times table we add 9 by adding 1 to the tens and subtracting 1 from the units. Therefore the sum of the digits stays the same. What happens if we continue the table beyond 10·9, what is the sum of the digits of the sum of the digits?

With two digits, each result is divisible by 9 and the digits of each result always sum to 9.

$$42 = 10·4 + 2 \text{ and } 24 = 10·2 + 4$$
$$42 - 24 = (10·4 + 2) - (10·2 + 4)$$
$$= 9·4 - 9·2$$
$$= 9(4 - 2)$$

Hence $42 - 24$ is divisible by 9. In general when we consider $ab - ba$ we are really saying $(10a + b) - (10b + a)$, which is $9a - 9b$ or $9(a - b)$. Thus it is clear that 9 will always be a divisor of the result. To see that the sum of the digits is always 9, we again consider $ab - ba$ and notice that since $a > b$ the unit column will contain $10 + b - a$, so now in the tens' column we will get $a - b - 1$. Hence the sum of the digits is $10 + b - a + a - b - 1 = 9$!

A similar thing happens with three digit numbers:

$$(100a + 10b + c) - (100c + 10b + a) = 100(a - 1) + 10(10 + b - 1) + 10 + c$$
$$- (100c + 10b + a)$$
$$= 100(a - 1 - c) + 90 + 10 + c - a$$
$$= 99a - 99c$$
$$= 99(a - c)$$

So 99 is always a divisor of the result. The digits of the answer are $a - c - 1$, 9, $10 + c - a$ which if we add them together give $19 - 1 = 18$. Hence the sum of the digits will always be 18.

Since $p = a - c - 1$, $q = 9$, $r = 10 + c - a$ and hence

$$pqr + rqp = 100(a - c - 1) + 10(9) + 10 + c - a + 100(10 + c - a) + 10(9) + a - c - 1$$
$$= 100(9) + 180 + 9 = 1089$$

$$2332 = 2002 + 330 = 2·1001 + 3·110 = 2·11·91 + 3·11·10 = 11(2·91 + 3·10)$$

We can write our typical four digit palindromic number as $abba$:

$$abba = a(1001) + b(110) = a(11·91) + b(11·10) = 11(a·91 + b·10)$$

Hence 11 is always a divisor.

The earliest examples of our present number system were found on stone columns in India dating from about 250 BC and the positional notation can be found in a book of AD 825 by al-Khowarizmi, a Persian mathematician from whose name comes the word algorithm. His ideas were spread in Europe in the twelfth century when a Latin translation of his book was made.

2c Back to basics – comments

$$17 = 16 + 1 = 2^4 + 1$$
$$31 \cdot 17 = 31(16 + 1)$$
$$= 31 \cdot 16 + 31$$
$$= (31 \cdot 2) \cdot 8 + 31$$
$$= ((31 \cdot 2) \cdot 2) \cdot 4 + 31$$
$$= (((31 \cdot 2) \cdot 2) \cdot 2) \cdot 2 + 31$$

and thus multiplying 31 by 17 corresponds to doubling the number 31 four times and adding the number undoubled. This method will always work because any number can be written as a sum of powers of 2 and hence in order to multiply we only have to sum the right number of doubles of the multiplicand. We can write 17 in binary notation as 10001 and so multiplying 31 by 17 is the same as multiplying 31 by 2^4 and then adding 31. $(31 = 31 \cdot 2^0.)$

To see why we add the numbers which are opposite an odd number we need to look more closely at how to express a number in binary notation.

17 is odd, so the last digit of 17 when written in binary notation is 1. Remove that last digit of binary 17 and divide by 2 to obtain the preceding digits. 8 is even, so the last digit of 8 when written in binary notation is 0 and the last two digits of 17 when written in binary notation are 01. Remove the last digit of binary 8, namely 0, and divide by 2 to obtain the preceding digits. 4 is even, so the last digit of 4 when written in binary notation is 0 and the last three digits of 17 when written in binary notation are 001. Remove the last digit of binary 4, namely 0, and divide by 2 to obtain the preceding digits. 2 is even, so the last digit of 2 when written in binary notation is 0 and the last four digits of 17 when written in binary notation are 0001. Remove the last digit of binary 2, namely 0, and divide by 2 to obtain the preceding digits. 1 is odd, so the last digit of 1 when written in binary notation is 1 and the last five digits of 17 when written in binary notation are 10001.

Notice that the *ones* in the binary number only occur when the result of dividing by 2 is odd.

	"Halving"		Binary number	"Doubling"		
odd	17	=	10001	31		
	8	=	1000	62	=	$2^1 \cdot 31$
	4	=	100	124	=	$2^2 \cdot 31$
	2	=	10	248	=	$2^3 \cdot 31$
odd	1	=	1	496	=	$2^4 \cdot 31$

Binary $10001 = 2^4 + 1$. So $17 \cdot 31 = (2^4 + 1) \cdot 31 = 2^4 \cdot 31 + 1 \cdot 31 = 496 + 31$.

Would a similar method work if we took a third each time in the first column and trebled in the second column? Can any number be written as a sum of powers of 3? Are there any other series that allow any number to be written as a sum of different terms in them? For another possibility see the Fibonacci example in **20. Dominoes**.

To construct the cards we first write the numbers $1 = 2^0$, $2 = 2^1$, $4 = 2^2$, $8 = 2^3$, $16 = 2^4$, $32 = 2^5$ in each top left corner of the cards. Now we express each number from 1 to 63 in binary notation. 27 in binary notation is 11011 and so we write 27 on all the cards labelled $2^4 = 10000$, $2^3 = 1000$, $2^1 = 10$ and $2^0 = 1$.

The earliest example of this type of multiplication was found on the Rhind Papyrus of about 1650 BC. The papyrus gives us a mathematical snapshot of the ancient Egyptians. It shows how they performed both multiplication and division by a sequence of doubling operations. It is believed that one of the problems on the papyrus is an early version of the children's rhyme "As I was going to St Ives, I met a man with seven wives, the seven wives had seven sacks, the seven sacks had seven cats, the seven cats had seven kits. How many were going to St Ives?"

For more information on the Rhind Papyrus see *Introduction to the History of Mathematics* by H. Eves published by Holt, Rinehart and Winston, 1969, or *The Rhind mathematical papyrus* by A.B. Chace published by N.C.T.M., 1979.

3c Children's ages – comments

Primes are numbers different from 1 with exactly two divisors, themselves and 1. Any number can be factored uniquely into primes. For a proof of this see the notes on **9. A return visit to Algorithma**. 1 is not considered to be a prime number otherwise this would destroy the property of unique factorisation (into primes). The prime factors of 595 are 5, 7 and 17. Since the children's ages are primes we must both have three children aged 5, 7 and 17.

Be careful with the next one. The friend did not say that his children's ages were primes. If he had done so, he would either have made a mistake or else had a very unusual family, one child of 3 and quadruplets aged 2! The unique prime factorisation of 48 is $2 \cdot 2 \cdot 2 \cdot 2 \cdot 3$. So there are many possible answers. The ages could be $4(2 \cdot 2)$ and $12(2 \cdot 2 \cdot 3)$; 2 and 24; 2, 2 and 12; 2, 4 and 6; ... or even ages 1, 1, 4 and 12!

There have been many tables of factors of numbers produced. In 1659 the factors of numbers up to 24000 were printed as an appendix to a book on algebra. In the nineteenth century a group effort produced a 10 volume book with the factors of all numbers up to 10000000.

4c The prison door problem – comments

With 10 prison doors and 10 warders, only the doors numbered 1, 4 and 9 are left open. The numbers with exactly two factors are prime numbers. The only factors of the prime number p are 1 and p itself. The numbers with exactly three factors are the squares of prime numbers. If p is a prime number, the only factors of p^2 are 1, p and p^2. The factor lattices are for numbers of the form p, p^2, p^3, p^4, pq, p^2q, p^3q, p^2q^2, respectively, where p and q are distinct prime numbers.

2 prime	$21 = 3 \cdot 7$	41 prime	61 prime	$81 = 3^4$
3 prime	$22 = 2 \cdot 11$	$42 = 2 \cdot 3 \cdot 7$	$62 = 2 \cdot 31$	$82 = 2 \cdot 41$
$4 = 2^2$	23 prime	43 prime	$63 = 3^2 \cdot 7$	83 prime
5 prime	$24 = 2^3 \cdot 3$	$44 = 2^2 \cdot 11$	$64 = 2^6$	$84 = 2^2 \cdot 3 \cdot 7$
$6 = 2 \cdot 3$	$25 = 5^2$	$45 = 3^2 \cdot 5$	$65 = 5 \cdot 13$	$85 = 5 \cdot 17$
7 prime	$26 = 2 \cdot 13$	$46 = 2 \cdot 23$	$66 = 2 \cdot 3 \cdot 11$	$86 = 2 \cdot 43$
$8 = 2^3$	$27 = 3^3$	47 prime	67 prime	$87 = 3 \cdot 29$
$9 = 3^2$	$28 = 2^2 \cdot 7$	$48 = 2^4 \cdot 3$	$68 = 2^2 \cdot 17$	$88 = 2^3 \cdot 11$
$10 = 2 \cdot 5$	29 prime	$49 = 7^2$	$69 = 3 \cdot 23$	89 prime
11 prime	$30 = 2 \cdot 3 \cdot 5$	$50 = 2 \cdot 5^2$	$70 = 2 \cdot 5 \cdot 7$	$90 = 2 \cdot 3^2 \cdot 5$
$12 = 2^2 \cdot 3$	31 prime	$51 = 3 \cdot 17$	71 prime	$91 = 7 \cdot 13$
13 prime	$32 = 2^5$	$52 = 2^2 \cdot 13$	$72 = 2^3 \cdot 3^2$	$92 = 2^2 \cdot 23$
$14 = 2 \cdot 7$	$33 = 3 \cdot 11$	53 prime	73 prime	$93 = 3 \cdot 31$
$15 = 3 \cdot 5$	$34 = 2 \cdot 17$	$54 = 2 \cdot 3^3$	$74 = 2 \cdot 37$	$94 = 2 \cdot 47$
$16 = 2^4$	$35 = 5 \cdot 7$	$55 = 5 \cdot 11$	$75 = 3 \cdot 5^2$	$95 = 5 \cdot 19$
17 prime	$36 = 2^2 \cdot 3^2$	$56 = 2^3 \cdot 7$	$76 = 2^2 \cdot 19$	$96 = 2^5 \cdot 3$
$18 = 2 \cdot 3^2$	37 prime	$57 = 3 \cdot 19$	$77 = 7 \cdot 11$	97 prime
19 prime	$38 = 2 \cdot 19$	$58 = 2 \cdot 29$	$78 = 2 \cdot 3 \cdot 13$	$98 = 2 \cdot 7^2$
$20 = 2^2 \cdot 5$	$39 = 3 \cdot 13$	59 prime	79 prime	$99 = 3^2 \cdot 11$
	$40 = 2^3 \cdot 5$	$60 = 2^2 \cdot 3 \cdot 5$	$80 = 2^4 \cdot 5$	$100 = 2^2 \cdot 5^2$

The **fundamental theorem of arithmetic** says that every whole number greater than 1 can be expressed as a product of prime numbers in exactly one way. The proof of this is given in the notes on **9. A return visit to Algorithma**. $5359375 = 5^6 \cdot 7^3$. Any factor of 5359375 has the form $5^a \cdot 7^b$ where $a = 0, 1, 2, 3, 4, 5$ or 6, and $b = 0, 1, 2$ or 3. So with seven possibilities for a and four possibilities for b, there are 28 factors in all.

2^x has $x + 1$ factors.
$2^x \cdot 3^y$ has $(x + 1)(y + 1)$ factors.
$2^x \cdot 3^y \cdot 5^z$ has $(x + 1)(y + 1)(z + 1)$ factors.
$2^x \cdot 3^y \cdot 5^z \cdot 7^t$ has $(x + 1)(y + 1)(z + 1)(t + 1)$ factors.
$2^x \cdot 3^y \cdot 5^z \cdot 7^t$ has an odd number of factors if $(x + 1)(y + 1)(z + 1)(t + 1)$ is odd, which is so if and only if all its factors are odd, that is if and only if x, y, z and t are all even and the number is a square.

When you think you can see what the fundamental theorem of arithmetic is saying, you may like to use it to sort out which numbers are *square-free*.

The number 30 is *square-free* because none of its divisors, 1, 2, 3, 5, 6, 10, 15 and 30, except 1, is a square number. The number 28 is *not square-free* because the square number 4 is one of its divisors. The numbers 27 and 45 are *not square-free* because the square number 9 is one of their divisors.

The fact that a product of k distinct prime numbers has 2^k factors was known to Cardan (1537) and the general formula $(a + 1)(b + 1)(c + 1) \ldots$ for the number of factors of $p_1^a p_2^b p_3^c \ldots$ was first published by John Kersey (1673).

5c Catching practice – comments

With 12 cricketers in the circle, steps of 1, 5, 7 and 11 reach every cricketer in the circle while steps of 2, 3, 4, 6, 8, 9, 10 do not. With nine cricketers in the circle, steps of 1, 2, 4, 5, 7 and 8 reach every cricketer in the circle, while steps of 3 and 6 do not. The critical number is the highest common factor of the step length and the number in the circle. If the highest common factor is 1, every person is reached. If the highest common factor is more than 1, some people are not reached.

There are many interesting patterns which emerge. For example, in a circle of n cricketers, the path of the ball with a step length of a is mirrored by the path of the ball with a step length of $n - a$.

If clockwise steps are counted as positive and anti-clockwise steps as negative, we can arrange all the possible steps for the 12 person circle in such a way that each column corresponds to a person:

```
 . . .  . . .   . . .   . . .   . . .   . . .   . . .   . . .   . . .   . . .   . . .
-36 -35 -34 -33 -32 -31 -30 -29 -28 -27 -26 -25
-24 -23 -22 -21 -20 -19 -18 -17 -16 -15 -14 -13
-12 -11 -10  -9  -8  -7  -6  -5  -4  -3  -2  -1
  0   1   2   3   4   5   6   7   8   9  10  11  ← remainders
 12  13  14  15  16  17  18  19  20  21  22  23
 24  25  26  27  28  29  30  31  32  33  34  35
 . . .  . . .   . . .   . . .   . . .   . . .   . . .   . . .   . . .   . . .   . . .
```

Now every integer will appear exactly once in this array. The numbers in the left hand column are the multiples of 12, and are therefore of the form $12k$. Moving one step to the right in this array comes from adding 1. So every integer is of the form $12k + r$, where $r = 0$ (if the number is in the first column), 1 (if the number is in the second column), 2 (if the number is in the third column), . . . , 11 (if the number is in the last column).

There is nothing special about the choice of the number 12 in this. For the second circle of cricketers, nine columns would exhibit the whole story.

Generally if the integers are displayed in b columns, then any integer $a = bk + r$ where $0 \le r < b$. The number r is the remainder when a is divided by b. When a and b have been given, both k (which row is a in?) and r (which column is a in?) are unique. Insisting that r is positive produces slightly unexpected results when a is negative. For example, the remainder when -16 is divided by 12 is $+8$. The claim that for a given positive integer b, any integer a may be expressed uniquely as $bk + r$ with $0 \le r < b$ is known as the **division algorithm**.

6c Algorithma – comments

Whole numbers of fish may be exchanged for 12, 24, 36, 48, . . . sticks of sugar cane. A stick of sugar cane with 1, 2, 3, 4, . . . coconuts is worth 6, 11, 16, 21, . . . sticks of sugar cane. These numbers are in the first column of the table. The two sequences first meet at 36.

The crucial step from $[a, b]$ to $[a - b, b]$ when $a > b$ looks simple enough, and we can prove that the two pairs share the same common factor. Remember that all the numbers which occur here are positive whole numbers. If $d \mid a$ (which means d is a factor of a, and is usually read "d divides a") and $d \mid b$, then $a = d \cdot a'$ and $b = d \cdot b'$, so $[a, b] = [d \cdot a', d \cdot b']$ and $[a - b, b] = [d(a' - b'), d \cdot b']$. This proves that any common factor of $[a, b]$ is a common factor of $[a - b, b]$. Conversely, let $d \mid a - b = c$ (say) and $d \mid b$. Then $c = d \cdot c'$ and $b = d \cdot b'$, so $[a - b, b] = [c, b] = [d \cdot c', d \cdot b']$ and $[a, b] = [c + b, b] = [d \cdot c' + d \cdot b', d \cdot b'] = [d(c' + b'), d \cdot b']$. This proves that any common factor of $[a - b, b]$ is a common factor of $[a, b]$. We have completed the proof that $[a, b]$ and $[a - b, b]$ share the same highest common factor.

Since the argument may be repeated, the same is true all the way down the sequence $[a, b] \rightarrow [a - b, b] \rightarrow \ldots \rightarrow [c, c]$. The highest common factor of $[c, c]$ is obviously c, so c is the highest common factor of $[a, b]$.

Now every step down the sequence is taken by subtracting as and bs from as and bs, so every number which occurs is of the form $xa + yb$ for some integers x and y (positive or negative).

The final conclusion is that the highest common factor of a and b, hcf(a, b), may be expressed in the form $xa + yb$.

The process of repeating the step $[a, b]$ to $[a - b, b]$, always taking the lesser number from the greater, until the same number appears twice, is an automatic process for finding the highest common factor of two positive whole numbers. This process is known as **the Euclidean algorithm** since it appears as proposition 1 in *Euclid, Book VII* (c. 300 BC). Now you see where the island got its name. The **division algorithm** described in **section 5** isolates the number of times b may be subtracted from a in the Euclidean algorithm without reaching a negative number.

7c Breeding rabbits – comments

The total number of rabbit pairs each month comes from adding the adult pairs to the new-born pairs. The new-born pairs come from the total number of pairs two months back and the number of adult pairs comes from adding the number of adult pairs and the number of new-born pairs in the previous month. Hence each month's total is obtained adding the previous two months' totals and since the initial totals are 2 and 3 we obtain the Fibonacci sequence.

Month	Adult pairs	New-born pairs	Total pairs
1	1	1	2
2	2	1	3
3	3	2	5
4	5	3	8
5	8	5	13
6	13	8	21

The sequence continues 13, 21, 34, ... Each term is obtained by adding the previous terms.

Each term of the sequence depends on the previous two so in particular we have:

$$f(1) + f(2) = f(3) \qquad f(2) + f(3) = f(4)$$

Every pair of consecutive Fibonacci numbers can be shown to have a highest common factor of 1.

We know that $f(n + 1) = f(n) + f(n - 1)$ and suppose that d divides $f(n + 1)$ and $f(n)$; then it must divide $f(n - 1)$. Also $f(n) = f(n - 1) + f(n - 2)$ and thus since d divides $f(n)$ and $f(n - 1)$ then it must divide $f(n - 2)$ and so on until we get to d divides $f(1)$ but $f(1)$ is 1 and so d must also be 1.

Here are some examples of highest common factors of Fibonacci numbers alongside their Fibonacci representation:

$$\text{hcf}(610, 144) = 2, \text{ i.e. hcf}(f(15), f(12)) = f(3)$$
$$\text{hcf}(144, 121393) = 1, \text{ i.e. hcf}(f(12), f(26)) = f(2)$$
$$\text{hcf}(610, 832040) = 610, \text{ i.e. hcf}(f(15), f(30)) = f(15)$$
$$\text{hcf}(46368, 832040) = 8, \text{ i.e. hcf}(f(24), f(30)) = f(6)$$

You may have noticed that the highest common factor is always the same as the Fibonacci number of the highest common factor of their two Fibonacci representations. The following result states this more formally: the highest common factor of $(f(n), f(m))$ is $f(d)$ where d is the highest common factor of n and m. A proof of this can be found in *Elementary Number Theory* by D.M. Burton, Allyn and Bacon, 1980 (291–292).

For further information on properties of Fibonacci numbers see either *Fibonacci and Lucas Numbers*, by S. Vajda, Ellis Horwood, 1990, or *The Fibonacci Quarterly*, Fibonacci Association, Santa Clara University, Santa Clara, CA 95053, USA.

8c To divide or not to divide – comments

You will have discovered that prime divisors of a product will always divide one of the factors, but composite divisors of a product will not necessarily do so. We can express this formally as:

Lemma: Let p be a prime number and a and b be integers. If p divides ab then p divides a, or p divides b.

This little result is also useful when you wish to prove the result that you used in **3. Children's ages**, that every number has a unique decomposition into prime factors. For more about this, see **9. A return visit to Algorithma**.

9c   A return visit to Algorithma – comments

The crucial step is taken after reaching the equation $(x - 5) \cdot 5 + (y + 2) \cdot 12 = 0$. Then $(y + 2) \cdot 12 = (5 - x) \cdot 5$ and since 5 divides the right hand side, 5 must divide the left hand side. The number 5 is prime so if 5 divides $(y + 2) \cdot 12$, *either* 5 divides $(y + 2)$ *or* 5 divides 12 as in **section 8**.

We justify this argument with a **lemma** (small theorem) which can be deduced from the Euclidean algorithm, that the highest common factor of $[a, b]$ may be expressed in the form $xa + yb$.

A prime number is a number ($\neq 1$) whose only (two) factors are itself and 1. We seek to prove (**the lemma**) that **if a prime number p divides a product $a \cdot b$, then either $p \mid a$, or $p \mid b$** (or possibly both). For example, $24 = 2 \cdot 12 = 3 \cdot 8 = 4 \cdot 6$, and for each product at least one of the components has a factor 2 and one has a factor 3.

To prove this lemma, let us suppose that $p \mid a \cdot b$, but that p does not divide a. As p only has 1 and p as factors, the highest common factor of p and a is 1. The Euclidean algorithm now tells us that there are integers x and y such that $xp + ya = 1$. Multiplying both sides of this equation by b gives $xpb + yab = b$. As both of the terms on the left hand side are divisible by p, p is a factor of the left hand side. Thus $p \mid b$ and our proof is complete.

Now we return to the original problem. Since $y + 2$ is a multiple of 5, $y + 2 = 5t$. So $5t \cdot 12 = (5 - x) \cdot 5$ and $x = -12 \cdot t + 5$.

It is quite easy to repeat the argument of the lemma to show that **if c divides the product $a \cdot b$ and c and a have no common factor greater than 1, then c divides b.** For $\mathrm{hcf}(a, c) = 1 \Rightarrow$ the existence of integers x and y such that $xa + yc = 1$. Multiplying by b gives $xab + ycb = b$. Now both the terms on the left hand side are divisible by c, so the right hand side must be too, and $c \mid b$.

Now for given integers a, b and c, we have all the ideas needed to decide whether there is an integer solution to the equation $xa + yb = c$, that is whether c is a multiple of $\mathrm{hcf}(a, b)$ or not, and to show that if we can find one solution we can find an infinity of solutions.

The lemma that if $p \mid a \cdot b$ then $p \mid a$ or $p \mid b$ is also the essential building brick for the proof of the fundamental theorem of arithmetic, first mentioned in **section 3**.

The **fundamental theorem of arithmetic** states that if a whole number greater than 1 is expressed as a product of prime numbers, then this expression is unique except for the order in which the prime numbers appear in the product.

To see how the proof works, suppose that p_1, p_2, q_1 and q_2 are prime numbers and that some number N is equal to the product $p_1 \cdot p_2$ and also to the product $q_1 \cdot q_2$. That is, $N = p_1 \cdot p_2 = q_1 \cdot q_2$. Now $p_1 \mid N$ so $p_1 \mid q_1 \cdot q_2$, and so $p_1 \mid q_1$ or $p_1 \mid q_2$, but q_1 and q_2 are primes, so either $p_1 = q_1$ (leaving $p_2 = q_2$) or $p_2 = q_1$ (leaving $p_1 = q_2$).

Now for the general proof. Let us suppose that $p_1, \ldots, p_n, q_1, \ldots, q_m$ are prime numbers, and $p_1 \cdot p_2 \cdot p_3 \cdot \ldots \cdot p_n = q_1 \cdot q_2 \cdot q_3 \cdot \ldots \cdot q_m$. Since p_1 is a prime number which divides the first product, it must divide the second product. So, by the lemma, either $p_1 \mid q_1$ or $p_1 \mid q_2 \cdot q_3 \cdot \ldots \cdot q_m$. If the first possibility holds, $p_1 = q_1$, and if the second possibility holds then the same kind of argument may be used, up to m times if necessary, to show that p_1 is equal to one of the q_i. Without loss of generality, we suppose that $p_1 = q_1$. Now we can repeat the argument with the equation $p_2 \cdot p_3 \cdot \ldots \cdot p_n = q_2 \cdot q_3 \cdot \ldots \cdot q_m$. And one by one we can prove that the primes on each side of the equation are equal in pairs.

10c The stamp problem – comments

Using 5p stamps alone, or 7p stamps alone, the following numbers (shown in bold face) can be reached:

1	2	3	4	**5**
6	**7**	**8**	**9**	**10**
11	**12**	**13**	**14**	**15**
16	**17**	**18**	**19**	**20**
21	**22**	**23**	**24**	**25**
26	**27**	**28**	**29**	**30**
31	**32**	**33**	**34**	**35**
36	**37**	**38**	**39**	**40**

The multiples of 7 (7, 14, 21, 28, 35) all lie in different columns because if two were to lie in the same column, their difference would have to be a multiple of 5, and $7a - 7b = 7(a - b)$ is only a multiple of 5 when $a - b$ is. So if we can use both 7s and 5s in any quantity we can reach all the numbers in bold face:

1	2	3	4	**5**
6	**7**	**8**	**9**	**10**
11	**12**	**13**	**14**	**15**
16	**17**	**18**	**19**	**20**
21	**22**	**23**	**24**	**25**
26	**27**	**28**	**29**	**30**

31	**32**	**33**	**34**	**35**
36	**37**	**38**	**39**	**40**

If you start with 6p and 10p stamps the reachable numbers, in bold face, are

1	2	3	4	5	**6**
7	**8**	9	**10**	11	**12**
13	**14**	15	**16**	17	**18**
19	**20**	21	**22**	23	**24**
25	**26**	27	**28**	29	**30**
31	**32**	33	**34**	35	**36**
37	**38**	39	**40**	41	**42**
43	**44**	45	**46**	47	**48**
49	**50**	51	**52**	53	**54**
55	**56**	57	**58**	59	**60**

and no odd number can ever be reached because $6x + 10y$ is always even. The crux of the matter is that hcf(6, 10) = 2. But if hcf(a, b) = 1, then the argument for 5p and 7p stamps can be carried through.

The reachable numbers in the second array are all of the form $7x + 5y$ where x and y are integers but not negative. The unreachable numbers are all above 7, 14, 21 or 28 and so are of the form $7x + 5y$ where $x = 1, 2, 3$ or 4 and y is *negative*.

If u is unreachable and $u = 7x + 5y$, then $30 - u = 7(5 - x) + 5(-1 - y)$ where the coefficients of 7 and 5 are not negative, and so $30 - u$ is reachable. Look at the numbers inside the box in the second display. Conversely if u is reachable, $30 - u$ is unreachable, so the non-multiples of 5 below 28 are in (reachable, unreachable) pairs, so there must be 14 of each.

A different approach to the postage stamp problem is provided in chapters 12 to 15 of *Discovering Mathematics* by A. Gardiner, Oxford University Press, 1987.

11c   Tests for divisibility – comment

(i)   An integer is divisible by 10 if its last digit is 0. $abcd0 = 10 \cdot abcd$.
(ii)   An integer is divisible by 5 if its last digit is 0 or 5. $abcd5 = 5(2 \cdot abcd + 1)$.
(iii) An integer is divisible by 2 if its last digit is even. For example, $abcd6 = 2(5 \cdot abcd + 3)$.

$$abc = 100a + 10b + c = 99a + 9b + (a + b + c)$$

Thus abc is divisible by 9 when $a + b + c$ is divisible by 9.

This rule was certainly known by AD 1000.

And abc is divisible by 3 when $a + b + c$ is divisible by 3.

$$abcdefg = 1000000a + 100000b + 10000c + 1000d + 100e + 10f + g$$
$$= 999999a + 99999b + 9999c + 999d + 99e + 9f + (a + b + c + d + e + f + g)$$

And $abcdefg$ is divisible by 3 when $a + b + c + d + e + f + g$ is divisible by 3.

abc is divisible by 6 if c is even and $a + b + c$ is divisible by 3, because a number is divisible by 6 if and only if it is divisible by 2 and by 3. Similarly $abcdefg$ is divisible by 6 if g is even and $a + b + c + d + e + f + g$ is divisible by 3.

$$abcabc = abc \cdot 1001 = abc \cdot 7 \cdot 11 \cdot 13$$
$$1001 = 990 + 11 = 11 \cdot 90 + 11$$
$$100001 = 99990 + 11 = 11 \cdot 9090 + 11$$
$$10000001 = 9999990 + 11 = 11 \cdot 909090 + 11$$
$$abcd = 1000a + 100b + 10c + d$$
$$= 1001a - a + 99b + b + 11c - c + d$$
$$= 1001a + 99b + 11c + (-a + b - c + d)$$

which is divisible by 11 if $-a + b - c + d$ is.

$$abcde = 10000a + 1000b + 100c + 10d + e$$
$$= 9999a + a + 1001b - b + 99c + c + 11d - d + e$$
$$= 9999a + 1001b + 99c + 11d + (a - b + c - d + e)$$

which is divisible by 11 if $a - b + c - d + e$ is.

This rule was certainly known by AD 1000.

Since $x \cdot 100$ is divisible by 4, abc is divisible by 4 if bc is: $bc = 10b + c = 8b + 2b + c$

So bc is divisible by 4 if $2b + c$ is. Since $x \cdot 1000$ is divisible by 8, $abcd$ is divisible by 8 if bcd is: $bcd = 100b + 10c + d = 96b + 8c + (4b + 2c + d)$

So bcd is divisible by 8 if $4b + 2c + d$ is.

$10 \cdot (abcd - 2 \cdot e) = abcde - 21 \cdot e$ so $abcde$ has a factor 7 if and only if $abcd - 2 \cdot e$ does.

Some tests for divisibility by 7 were given by Pisano in 1202.

13c Eratosthenes' sieve – comment

Eratosthenes (275–194 BC) devised the first sieve method for finding prime numbers. He began with a list of odd numbers: he deleted 32 and every third number thereafter, he deleted 52 and every fifth number thereafter, and so on. Arab mathematicians by the end of the thirteenth century had recognised that in order to find all primes less than N it is only necessary to delete multiples of primes less than $\sqrt{N}$.

The method used in **section 13** is based on the following special case. Every non-prime number less than 100 has a factor ($\neq 1$) which is less than 10.

The proof is by contradiction: that is, we suppose that the result is wrong and obtain a contradiction. The contradiction is resolved by the correctness of the original claim.

To suppose that the result is wrong is to suppose that we can have a product $a \cdot b$ < 100, with $a \geq 10$ and $b \geq 10$. Now $b \geq 10 \Rightarrow 10 \cdot b \geq 100$, and $a \geq 10 \Rightarrow a \cdot b \geq 10 \cdot b$ ≥ 100, so $a \cdot b \geq 100$, which contradicts the hypothesis.

Thus every composite number less than 100 has a factor less than 10. Now if N (< 100) is our composite number and d is a factor of N which is less than 10, d has a factor 2, 3, 5 or 7. Clearly a factor of d has to be a factor of N. So N has 2, 3, 5 or 7 as a factor. Thus if the multiples of 2, 3, 5 and 7 are deleted from the numbers between 2 and 100, only prime numbers remain.

Likewise if $a \geq 15$ and $b \geq 15$, then $a \cdot b \geq 225 > 200$, so every composite number less than 200 has a factor less than 15, and so a prime factor less than 15.

To express the result generally, **every composite number $\leq N$ has a factor $\leq$** $\sqrt{N}$.

There are 25 prime numbers between 1 and 100, but only 20 between 100 and 200. The prime numbers gradually become more sparse. In fact the number of primes less than n gets closer and closer to $n/\log_e n$ as n gets large, and there are very long runs of consecutive numbers containing no primes at all. Consider for example the list of numbers $n! + a$, where $a = 2, 3, 4, \ldots, n$, and see **section 14**.

$10n$, $10n + 2$, $10n + 4$, $10n + 6$ and $10n + 8$ each have a factor 2 and $10n + 5$ has a factor 5; so all prime numbers, apart from 2 and 5, have one of the forms $10n + 1$, $10n + 3$, $10n + 7$ and $10n + 9$, and there is an infinity of primes of each of these forms.

14c Raffle tickets and neighbours – comments

Composites

From the table, using Eratosthenes' sieve we can find runs $8, 9, 10$ and $24, 25, 26,$ $27, 28$ and $90, 91, 92, 93, 94, 95, 96$. Runs of consecutive non-primes: $26, 27, 28$ divisible by $2, 3, 4$; $122, 123, 124, 125$ divisible by $2, 3, 4, 5$.

In general if we take $n! + 2, n! + 3, \ldots, n! + n$ we will obtain a sequence of $n - 1$ consecutive non-primes.

Consecutive primes

The largest prime pair known to date is

$$1639494(2^{4423} - 1) - 1 \text{ and } 1639494(2^{4423} - 1) + 1$$

discovered by Keller in 1983. "Is there an infinite number of prime pairs?" is an open question.

The sum of prime pairs is always divisible by 12. All primes after 2 and 3 are of the form $12n + 1, 12n + 5, 12n + 7$ or $12n + 11$. Prime pairs are of the form $12n + 11$ and $12(n + 1) + 1$ or $12n + 5$ and $12n + 7$. Thus their sum is of the form $24(n + 1)$ or $24n + 12$ which are both divisible by 12.

If we take the product of the primes of a prime pair plus 1 the answer is always a perfect square. In fact for any two integers that are 2 apart we have $a(a + 2) = (a + 1)^2 - 1$.

$3, 5, 7$ is the only prime triple. Let our prime triple be of the form $p, p + 2, p + 4$. The number 3 will divide one of any three consecutive numbers so in particular 3 must divide one of $p, p + 1, p + 2$. If we assume p is greater than 3, then 3 does not divide p or $p + 2$ so 3 must divide $p + 1$ and hence will also divide $(p + 1) + 3 = p + 4$. Thus $p + 4$ is not a prime. Thus p is at most 3 and the only prime triple is $3, 5, 7$.

Hence James's house is number 5. This also follows from the result above that all primes after 2 and 3 are of the form $12n + 1, 12n + 5, 12n + 7$ or $12n + 11$.

For further information see the article *Runs of composite integers and the Chinese remainder theorem* by Rex Watson in *Mathematical Gazette*, Vol. 78, No. 482, July 1994, pages 167–172.

15c How many primes? – comment

30031. The remainder must be 1.

$$174^2 = 30276 > 30031 > 29929 = 173^2$$

From its construction 30031 is not divisible by any prime number less than or equal to 13. If it is not prime, it must have a factor less than or equal to 173. Now use the list of primes from **13. Eratosthenes' sieve** to find the factor 59.

$p_1 \cdot p_2 \cdot \ldots \cdot p_n + 1$ is not divisible by any of the prime numbers $p_1, p_2, \ldots, p_n$.

We can prove that there is an infinity of primes by contradiction. Suppose that there are only a finite number (n) of primes. Let those primes be $p_1, p_2, \ldots, p_n$; then we have already shown how to construct a number which is not divisible by any of them. The prime factors of this number are not in the list. This contradicts the supposition that there was a determinate number of primes. So **there is an infinity of prime numbers**.

The argument here first appeared in Euclid, Book IX (c. 300 BC).

16c No shuffling – comments

Wendy	Toots	Tweedle	me
1	2	3	4
5	6	7	8
9	10	11	12
13	14	15	16
17	18	19	20
21	22	23	24
25	26	27	28
29	30	31	32
33	34	35	36
37	38	39	40
41	42	43	44
45	46	47	48
49	50	51	52
53	54	55	56
57	58	59	60
61	62	63	64
65	66	67	68
69	70	71	72
73	74	75	76
77	78	79	80
81	82	83	84
85	86	87	88
89	90	91	92
93	94	95	96
97	98	99	100

The numbers on all the cards in my hand are multiples of 4, so each is $4n$ for some whole number n. The numbers on all the cards in Wendy's hand leave remainder 1 when divided by 4, so each is $4n + 1$ for some whole number n. The numbers on Toots' cards all leave remainder 2 when divided by 4, so each is $4n + 2$ for some whole number n. The numbers on Tweedle's cards all leave remainder 3 when divided by 4, so each is $4n + 3$ for some whole number n.

Any two of Wendy's cards must have numbers of the form $4n + 1$ and $4m + 1$ on them. Their sum is $(4n + 1) + (4m + 1) = 4(n + m) + 2$, which is the number on one of Toots' cards. If we express this as Wendy + Wendy = Toots, then we similarly get other sums as in this table:

+	Wendy	Toots	Tweedle	me
Wendy	Toots	Tweedle	me	Wendy
Toots	Tweedle	me	Wendy	Toots
Tweedle	me	Wendy	Toots	Tweedle
me	Wendy	Toots	Tweedle	me

If we multiply the numbers on two of Tweedle's cards together we get

$$(4n + 3)(4m + 3) = 16mn + 12n + 12m + 9$$
$$= 4(4mn + 3m + 3n + 2) + 1$$

which is the number on one of Wendy's cards. If this is expressed by saying Tweedle·Tweedle = Wendy, then we similarly get other products as in this table:

	Wendy	Toots	Tweedle	me
Wendy	Wendy	Toots	Tweedle	me
Toots	Toots	me	Toots	me
Tweedle	Tweedle	Toots	Wendy	me
me	me	me	me	me

Only Wendy and I hold cards with square numbers on them: look down the main diagonal of the product table.

I hold no primes because all my numbers have a factor 4. Toots holds 2 but no other primes because all Toots' numbers have a factor 2. Wendy and Tweedle each hold many primes.

17c Can you really tell the time? – comments

The minute hand

The minute hand points upwards when the number of quarter-hours is a multiple of 4. The minute hand is at the first quarter when the number of quarter-hours is (a multiple of 4) + 1. The minute hand is at the half-hour when the number of quarter-hours is (a multiple of 4) + 2. The minute hand is at the third quarter when the number of quarter-hours is (a multiple of 4) + 3.

±100 are both in the first column; 99 is in the fourth column; −99 is in the second column.

Numbers in the same column are separated by a multiple of 4..
$6 \equiv -6 \pmod 4$, 23 is not $\equiv -23 \pmod 4$, $300 \equiv 500 \pmod 4$, $17 \equiv 5 \pmod 4$, 17 is not $\equiv -5 \pmod 4$, 16 is not $\equiv -18 \pmod 4$.

The hour hand

Moving on or back a multiple of 12 hours gives the same position of the hour hand. 9 is not $\equiv 3 \pmod{12}$, $9 \equiv -3 \pmod{12}$, $100 \equiv 4 \pmod{12}$, 100 is not $\equiv -4 \pmod{12}$, 45 is not $\equiv 7 \pmod{12}$, $87 \equiv 15 \pmod{12}$.

> $a \equiv b \pmod 4$ means
> a and b have the same remainder when divided by 4,
> or $a - b$ is a multiple of 4,
> or $b - a$ is a multiple of 4,
> or 4 is a factor of $a - b$,
> or 4 divides $a - b$, written $4 \mid a - b$,
> or $a - b = 4k$ for some integer k,
> or $a = 4k + b$.

> $a \equiv b \pmod 4$ means a and b have the same remainder when divided by 12, or $a - b$ is a multiple of 12, or $b - a$ is a multiple of 12, or 12 is a factor of (or divides) $a - b$, written $12 \mid a - b$, or $a - b = 12k$ for some integer k, or $a = 12k + b$.

Overview

Telling the time used two displays for the integers, one with four columns and one with 12.

The general property that, having chosen a positive whole number b, all the integers may be displayed in b columns with a central row reading from 0 to $b - 1$, with each row below being obtained by repeatedly adding b and each row above being obtained by repeatedly subtracting b, can be described by the **division algorithm**, given in the comments on **5. Catching practice**. Any integer a appears just once in the display. If a appears in the qth row below the central row, $a = bq + r$, where r, which specifies the column, is an entry in the central row, that is $0 \le r < b$. For example, taking $a = 100$ and $b = 12$, $100 = 12 \cdot 8 + 4$. The number q ($= 8$ in this case) is the quotient and the number r ($= 4$ in this case) is the remainder when a is divided by b. In particular $a \equiv r \pmod b$. The letters q and r are the initial letters of *quotient* and *remainder*, the numbers that result when we attempt to divide a by b.

The fact that every integer lies in just one of the four columns (mod 4) or just one of the 12 columns (mod 12), or generally in just one of the n columns (mod n), follows from the fact that the relation "$\equiv$" on the integers is

(i)   *reflexive*, which means $a \equiv a \pmod n$ for all integers a;
(ii) *symmetric*, which means that when $a \equiv b \pmod n$, $b \equiv a \pmod n$;
(iii) *transitive*, which means that when $a \equiv b$ and $b \equiv c$, then $a \equiv c$.

The reflexive and symmetric properties are obvious. The transitive property comes from adding the equations $a - b = kn$ and $b - c = ln$. The conventional symbol to denote the set of integers is **Z**. The relation "≡" groups the integers into columns of congruent integers, each integer belonging to exactly one group (or column). The conventional symbol to denote the set of columns of integers when the integers have been displayed in n columns is $\mathbf{Z}_n$.

The meaning and significance of the terms *reflexive, symmetric, transitive* and *relation* are elaborated in **19. Relations can be difficult.**

All the integers congruent to a given integer (i.e. one column of the array *modulo n*) have the same remainder on division by n and such a set is called a residue class *modulo n*.

Now the miniature addition table *modulo 4*.

$$(\mathbf{Z}_4, +)$$

+	0	1	2	3
0	0	1	2	3
1	1	2	3	0
2	2	3	0	1
3	3	0	1	2

takes on a new perspective, because each number here stands for a whole column of integers. The claim is that (for example) any number in the 1 column of the original display added to any number in the 2 column always results in a number in the $1 + 2 = 3$ column. More generally, if $a \equiv b \pmod{n}$ and $c \equiv d \pmod{n}$, then $a + c \equiv b + d \pmod{n}$, which follows from adding $a - b = kn$ to $c - d = ln$. This is the justification for claiming that addition works consistently on the columns of numbers, and that there is such a thing as $(\mathbf{Z}_n, +)$.

18c Fibonacci numbers and the division algorithm – comments

The division algorithm says that for any integers a and $b > 0$ there exist unique integers q and r so that

$$a = bq + r \text{ and } 0 \le r < b$$

$f(7) = 13, f(6) = 8$, so using the division algorithm

$$\begin{aligned}
13 &= 1 \cdot 8 + 5 \\
8 &= 1 \cdot 5 + 3 \\
5 &= 1 \cdot 3 + 2 \\
3 &= 1 \cdot 2 + 1 \\
2 &= 2 \cdot 1 + 0
\end{aligned}$$

Hence five applications of the division algorithm are required to obtain hcf($f(7)$, $f(6)$). Similarly it takes six applications to obtain hcf($f(8)$, $f(7)$). We might expect to obtain hcf($f(12), f(11)$) in 10 applications and in general to obtain hcf($f(n+2)$, $f(n+1)$) in n applications. The reason for this can be seen if we write the steps out in general:

$$f(n+2) = f(n+1) + f(n)$$
$$f(n+1) = f(n) + f(n-1)$$
$$f(n) \quad = f(n-1) + f(n-2)$$

$$\cdot$$
$$\cdot$$
$$\cdot$$

$$f(4) \quad = f(3) + f(2)$$
$$f(3) \quad = 2f(2) + 0$$

which is n applications of the division algorithm and the last non-zero remainder is $f(2)$ which is 1.

Alternatively using the method in **6. Algorithma**

$$[f(6), f(5)] = [f(4), f(5)] = [f(4), f(3)] = [f(3), f(2)] = [f(2), f(1)] = [1, 1]$$

19c Relations can be difficult – comments

Symmetric relations

If we take X as Jack then although Lisa is Jack's sister Jack is not Lisa's sister. However, when we take the relationship "is a sibling of" the relationship is symmetric.

▷ "has the same parents as" is symmetric.

▷ "is the father of" is not symmetric: Jonathan is the father of Jack but Jack is not the father of Jonathan.

▷ "is married to" is symmetric.

▷ "is an aunt of" is not symmetric: Betty is an aunt of Penny but Penny is not an aunt of Betty.

1. "is less than" for integers is not symmetric: $3 < 4$ but $4 \not< 3$.
2. "is the opposite of" for words in English is symmetric.
3. "is the same distance from the origin as" for points on a plane is symmetric.
4. "is a factor of" for integers is not symmetric: 3 is a factor of 12 but 12 is not a factor of 3.
5. "is congruent *modulo* 4 to" for integers is symmetric:

$$a \equiv b \,(\mathrm{mod}\ 4) \Rightarrow 4 \mid (a-b) \Rightarrow 4 \mid (b-a) \Rightarrow b \equiv a \,(\mathrm{mod}\ 4).$$

6. "is perpendicular to" for lines in a plane is symmetric.
7. "is parallel to" for lines in a plane is symmetric.

Transitive relations

1. "is less than" for integers is transitive.
2. "is the opposite of" for words in English is not transitive. For example, wet is the opposite of dry and dry is the opposite of wet but wet is not the opposite of wet.
3. "is the same distance from the origin as" for points on a plane is transitive.
4. "is a factor of" for integers is transitive.
5. "is congruent *modulo* 4 to" for integers is transitive:

$$a \equiv b(\text{mod } 4) \Rightarrow 4 \mid (a - b) \Rightarrow \quad a - b \quad = 4k \text{ for some integer } k$$
$$b \equiv c(\text{mod } 4) \Rightarrow 4 \mid (b - c) \Rightarrow \quad b - c \quad = 4m \text{ for some integer } m$$
$$\Rightarrow (a - b) + (b - c) = 4k + 4m = 4(k + m)$$
$$\Rightarrow 4 \mid (a - c)$$
$$\Rightarrow a \equiv c \text{ mod } 4$$

6. "is perpendicular to" for lines in a plane is not transitive.
7. "is parallel to" for lines in a plane is transitive.

Reflexive relations

Since no one is the mother of themselves this relation is not reflexive.

1. "is less than" for integers is not reflexive: no number is less than itself.
2. "is the opposite of" for words in English is not reflexive.
3. "is the same distance from the origin as" for points on a plane is reflexive.
4. "is a factor of" for integers is reflexive.
5. "is congruent *modulo* 4 to" for integers is reflexive:

$$a \equiv a \ (\text{mod } 4) \Leftrightarrow 4 \mid (a - a) = 0 = 4 \times 0.$$

6. "is perpendicular to" for lines in a plane is not reflexive.
7. "is parallel to" for lines in a plane is reflexive if a line is counted as parallel to itself.

The relation $a \equiv b \ (\text{mod } 5)$ partitions the set $\{0, 1, 2, 3, 4, 5, 6, 7, 8, 9, 10\}$ into classes $\{0, 5, 10\}, \{1, 6\}, \{2, 7\}, \{3, 8\}$ and $\{4, 9\}$.

Any *relation* which is symmetric, transitive and reflexive always partitions the set into classes, called equivalence classes.

We can prove that $a \equiv b \ (\text{mod } n)$ has all three properties:

1. To prove congruence is symmetric: that is, $a \equiv b \ (\text{mod } n) \Rightarrow b \equiv a \ (\text{mod } n)$.

   Proof:

$$a \equiv b \ (\text{mod } n) \Rightarrow n \mid (a - b) \Rightarrow n \mid (b - a) \Rightarrow b \equiv a \ (\text{mod } n)$$

2. To prove congruence is transitive:

   that is, $a \equiv b \ (\text{mod } n)$ and $b \equiv c \ (\text{mod } n) \Rightarrow a \equiv c \ (\text{mod } n).$

Proof:

$a \equiv b \pmod{n}$ and $b \equiv c \pmod{n} \Rightarrow n \mid (a - b)$ and $n \mid (b - c)$

$$\Rightarrow a - b = kn \text{ and } b - c = ln \text{ for some integers } k, l$$
$$\Rightarrow a - b + b - c = kn + ln$$
$$\Rightarrow a - c = (k + l)n$$
$$\Rightarrow n \mid a - c$$
$$\Rightarrow a \equiv c \pmod{n}$$

3. To prove congruence is reflexive: that is, $a \equiv a \pmod{n}$.

Proof:

$$a - a = 0n \Rightarrow n \mid (a - a) \Rightarrow a \equiv a \pmod{n}$$

Hence the relation $a \equiv b \pmod{n}$ has all three properties and the relation partitions the integers into related classes. All the integers in one class have the same remainder when divided by n. This is why the classes may be called *residue* classes.

Non-transitive dice

A/B	5	5	5	5	1	1
6	A	A	A	A	A	A
6	A	A	A	A	A	A
2	B	B	B	B	A	A
2	B	B	B	B	A	A
2	B	B	B	B	A	A
2	B	B	B	B	A	A

We would expect A to win over B on average $20/36$ times.

Similar tables for B and C give B to win over C on average $24/36$ times and C to win over A on average $24/36$ times.

20c Dominoes – comments

To know if the dominoes will fall we need to know that the first one falls and that if any one domino falls then it will topple the next one in the line.

Examples 1 and 3 are true and example 2 is false for $n \geq 7$.

To prove example 1 we add the next term $(n + 1)^2$ to both sides to get

$$1^2 + 2^2 + 3^2 + \ldots + n^2 + (n + 1)^2 = \tfrac{1}{6}n(n + 1)(2n + 1) + (n + 1)^2$$
$$= \tfrac{1}{6}(n + 1)[n(2n + 1) + 6(n + 1)]$$
$$= \tfrac{1}{6}(n + 1)(2n^2 + 7n + 6)$$
$$= \tfrac{1}{6}(n + 1)(n + 2)(2n + 3)$$
$$= \tfrac{1}{6}(n + 1)[(n + 1) + 1][2(n + 1) + 1]$$

To do example 3 we add the next term $(n + 1)$ to both sides to get

$$1 + 2 + 3 + \ldots + n + (n + 1) = \tfrac{1}{2}[n(n + 1)] + (n + 1)$$
$$= \tfrac{1}{2}(n + 1)(n + 2)$$

4. To prove $1^3 + 2^3 + 3^3 + \ldots + n^3 = [\tfrac{1}{2}n(n + 1)]^2$ for $n \geq 1$.

Let $S(n)$ mean $1^3 + 2^3 + 3^3 + \ldots + n^3 = [\tfrac{1}{2}n(n + 1)]^2$.

(i) $S(1)$ means $1^3 = [\tfrac{1}{2}1(1 + 1)]^2 = 1$, which is true.

(ii) If $S(n)$ is true then $1^3 + 2^3 + 3^3 + \ldots + n^3 = [\tfrac{1}{2}n(n + 1)]^2$ is true and by adding $(n + 1)^3$ to both sides we get

$$1^3 + 2^3 + 3^3 + \ldots + n^3 + (n + 1)^3 = [\tfrac{1}{2}n(n + 1)]^2 + (n + 1)^3$$
$$= \tfrac{1}{4}(n + 1)^2[(n^2 + 4(n + 1)]$$
$$= \tfrac{1}{4}(n + 1)^2(n + 2)^2$$
$$= [\tfrac{1}{2}(n + 1)(n + 2)]^2$$

and hence $S(n + 1)$ is true.

Thus by the principle of mathematical induction we have the required result.

5. To prove $6^n - 5n + 4$ is divisible by 5 for $n \geq 1$.

Let $S(n)$ mean $6^n - 5n + 4$ is divisible by 5.

(i) $S(1)$ means $6 - 5 + 4 = 5$ is divisible by 5, which is true.

(ii) If $S(n)$ is true then $6^n - 5n + 4$ is divisible by 5 is true and by changing n to $n + 1$ we get

$$6^{(n+1)} - 5(n + 1) + 4 = 6(6^n - 5n + 4) + 25n - 25$$
$$= 6(6^n - 5n + 4) + 5(5n - 5)$$

which is divisible by 5 since by assumption $(6^n - 5n + 4)$ is divisible by 5 and hence $S(n + 1)$ is true.

Thus by the principle of mathematical induction we have the required result.

6. To prove $1 + a + a^2 + a^3 + a^4 + \ldots + a^n = (a^{n+1} - 1)/(a - 1)$ for $n \geq 1$.

Let $S(n)$ mean $1 + a + a^2 + a^3 + a^4 + \ldots + a^n = (a^{n+1} - 1)/(a - 1)$.

(i) $S(1)$ means $1 + a = (a^{1+1} - 1)/(a - 1)$, which is true.

(ii) If $S(n)$ is true then

$$1 + a + a^2 + a^3 + a^4 + \ldots + a^n = (a^{n+1} - 1)/(a - 1)$$

is true and by adding a^{n+1} to both sides we get

$$1 + a + a^2 + a^3 + a^4 + \ldots + a^n + a^{n+1} = (a^{n+1} - 1)/(a - 1) + a^{n+1}$$
$$= (a^{n+1} - 1 + a^{n+2} - a^{n+1})/(a - 1)$$
$$= (a^{n+2} - 1)/(a - 1)$$

and hence $S(n + 1)$ is true.

Thus by the principle of mathematical induction we have the required result.

For the two examples illustrating that **both** parts of the induction are necessary we have:

7. Let $S(n)$ mean $2 + 4 + \ldots + 2n = (n + 2)(n - 1)$.

   (i) $S(1)$ means $2 = (1 + 2)(1 - 1)$, which is not true.
   (ii) If $S(n)$ were true then $2 + 4 + \ldots + 2n = (n + 2)(n - 1)$ would be true and by adding $2(n + 1)$ to both sides we would get

   $$2 + 4 + \ldots + 2n + 2(n + 1) = (n + 2)(n - 1) + 2(n + 1)$$
   $$= (n + 3)n = [(n + 1) + 2] [(n + 1) - 1]$$

   and hence $S(n + 1)$ would be true.

   Thus the inductive step is satisfied, but the basis for the induction fails and the result is generally false.

8. Let $S(n)$ mean $n^2 + n + 41$ is a prime number.

   (i) $S(1)$ means $1^2 + 1 + 41$ is prime, which is true since 43 is prime.
   (ii) If $S(n)$ is true then $n^2 + n + 41$ is prime and we wish to say that this implies that $(n + 1)^2 + (n + 1) + 41$ is prime. Unfortunately we cannot say this: even though $S(1)$, $S(2)$, $\ldots$, $S(39)$ are all true $S(40)$ and $S(41)$ are not. Hence no inductive step is possible.

   Thus the basis for the induction holds but the inductive step fails and hence the result is false.

Induction on Fibonacci numbers

To express 100 as a sum of Fibonacci numbers we find the largest Fibonacci number less than 100, which is 89. Now $100 - 89 = 11$, so now we have to express 11 as a sum of Fibonacci numbers, $11 = 8 + 3$, and finally we have $100 = 89 + 8 + 3$.

   We can express this result as: every positive integer can be written as the sum of distinct Fibonacci numbers.

Proof:

Let $S(n)$ mean all numbers less than $f(n)$ can be written as the sum of different numbers from $f(1), \ldots, f(n - 1)$ for $n \geq 3$.

   (i) $S(3)$ is true since $f(3) = 2$ and 1 can be written as $f(1)$.
   (ii) If $S(n)$ is true then to show that $S(n + 1)$ is true we only need to check that all numbers x between $f(n)$ and $f(n + 1)$ can be written as the sum of different numbers from the set $\{f(1), \ldots, f(n - 1)\}$. For such x, $x - f(n) > 0$ and

   $$x - f(n) < f(n + 1) - f(n)$$
   $$= f(n - 1)$$

   so by property $S(n)$, $x - f(n)$ can be expressed as a disjoint sum of numbers from $f(1), \ldots, f(n - 2)$. Hence x can be written as a disjoint sum of numbers from $f(1), \ldots, f(n - 1)$ as required.

Thus by the principle of mathematical induction we have the required result.

Subsequences

Let $S(n)$ mean that $A(n)$ is divisible by 2.

(i) $S(1)$ means $A(1)$ is divisible by 2 which is true since $A(1) = 2$.
(ii) If $S(n)$ is true then $A(n)$ is divisible by 2 and

$$
\begin{aligned}
A(n+1) - A(n) &= f(3n+3) - f(3n) \\
&= f(3n+2) + f(3n+1) - f(3n) \\
&= 2f(3n+1) + f(3n) - f(3n) \\
&= 2f(3n+1)
\end{aligned}
$$

Hence $A(n+1) = A(n) + 2f(3n+1)$ and so $A(n+1)$ is also divisible by 2.

Thus by the principle of mathematical induction we have shown that $A(n)$ is divisible by 2 for all $n = 1, 2, 3, \ldots$.

The sequence $B(n)$ starts 3, 21, 144, 987, 6765, 46368, ... Let $S(n)$ mean that $B(n)$ is divisible by 3.

(i) $S(1)$ means $B(1)$ is divisible by 3 which is true since $B(1) = 3$.
(ii) If $S(n)$ is true then $B(n)$ is divisible by 3 and

$$
B(n+1) - B(n) = 3f(4n+1) + f(4n) = 3f(4n+1) + B(n)
$$

Hence $B(n+1) = 2B(n) + 3f(4n+1)$ and so $B(n+1)$ is also divisible by 3.

Thus by the principle of mathematical induction we have shown that $B(n)$ is divisible by 3 for all $n = 1, 2, 3, \ldots$.

To obtain a subsequence all of whose terms are divisible by 5 we need to take every sixth term of the Fibonacci sequence.

One of the earliest uses of proof by mathematical induction was by Pascal in 1653 in proving properties about his arithmetical triangle.

PART II

21c Chinese remainders – comments

The first table should have been filled thus:

Remainder on dividing by 5

		0	1	2	3	4
Remainder	0	15	6	12	3	9
on dividing	1	10	1	7	13	4
by 3	2	5	11	2	8	14

with the north-west south-east lines inviting routine filling.

The original question asks for a simultaneous solution to

$x \equiv 2 \pmod 3$,
$x \equiv 3 \pmod 5$ and
$x \equiv 2 \pmod 7$.

The strategy suggested starts by looking for an $x \pmod{15}$ which satisfies the first two conditions. It then looks for a solution to $x \equiv 8 \pmod{15}$ and $x \equiv 2 \pmod 7$.

The unique solution $\pmod{105}$ depends firstly on the fact that 3 and 5 have no common factor, and secondly on the fact that 15 and 7 have no common factor. The general result is that, provided m and n have no common factor, there is a unique solution *modulo mn* to $x \equiv a \pmod m$ and $x \equiv b \pmod n$. Because there are m different remainders $\pmod m$ and n different remainders mod n, there are mn pairs (a, b). Do the mn numbers $1, 2, 3, \ldots , mn$ each fit exactly one of the pairs?

Either the matching is one to one, or there is both doubling and unmatched pairs. We can indicate the reason why the matching is one to one when m and n are coprime by looking at a special case with $m = 8$ and $n = 9$:

$$72 = 8 \cdot 9 \text{ and } \mathrm{hcf}(8, 9) = 1$$

Conjecture how many solutions there may be to the simultaneous congruences $x \equiv 3 \pmod 8$ and $x \equiv 5 \pmod 9$, with $1 \le x \le 72$.

Suppose there are two solutions x and y. Then $x \equiv y \pmod 8$ and $x \equiv y \pmod 9$. So $x - y \equiv 0 \pmod 8$ and $x - y \equiv 0 \pmod 9$. Thus $8 \mid x - y$ and $9 \mid x - y$. Because $\mathrm{hcf}(8, 9) = 1$, it follows that $72 \mid x - y$. However, $\mid x - y \mid < 72$, and this is impossible unless $x = y$. So there is at most one solution to a pair of equations like that.

Now each number $1, 2, \ldots , 72$ is congruent to $0, 1, \ldots$ or $7 \pmod 8$ and congruent to $0, 1, \ldots$ or $8 \pmod 9$. So there are 72 possible pairs and 72 numbers. Since no two numbers fit the same pair, each number fits exactly one pair.

Notice that precisely the same argument holds had the original simultaneous congruences been $x \equiv a \pmod 8$ and $x \equiv b \pmod 9$, so these congruences also have a unique solution between 1 and 72.

x may be found by filling an 8×9 rectangular array or by an algebraic analysis as follows:

$$x \equiv 3 \pmod 8 \Rightarrow x - 3 = 8k$$
$$x \equiv 5 \pmod 9 \Rightarrow 8k + 3 \equiv 5 \pmod 9 \Rightarrow 8k \equiv 2 \pmod 9$$

Multiplying by 8 gives $64k \equiv 7 \pmod 9$, or $k \equiv 7 \pmod 9$. $k = 7$ gives $x = 59$.

When two numbers m and n have no common factor, entering the numbers from 1 to mn in an $m \times n$ rectangular array can be done fitting one number into one position with the numbers in any column having the same remainder when divided by m and the numbers in any row having the same remainder when divided by n. The reason is that there are mn numbers and mn positions, so that if they did not fit one to one, there would be both an empty position (and there could be more than 1) and a position with more than one number in it.

So we pretend that we have two numbers between 1 and mn in the same position. If x and y are two numbers between 1 and mn, and both have remainder a when divided by m and remainder b when divided by n, then $x \equiv a \equiv y \pmod m$. This makes $x - y$ a multiple of m and so $x - y = km$.

Similarly, $x \equiv b \equiv y \pmod n$. This makes $x - y$ a multiple of n. So km is a multiple of n. However, m and n have no common factors, so k is a multiple of n (*: the proof is given in the comments on **9. A return visit to Algorithma**), and if $k = k'n$, then $x - y = km = k'mn$. So $x - y$ is a multiple of mn. But both x and y are between 1 and mn, so their difference is numerically less than mn. Since k' is an integer, $k' = 0$. Thus $x = y$.

This argument depends critically, at stage (*), on m and n having no common factors. Because 3 and 5 have no common factors, the first table can be built up. Because 15 and 7 have no common factors, the second table can be built up. Because 4 and 6 have the common factor 2, there are either two numbers below 24 satisfying the two conditions, or there are none at all. The rectangular 4×6 array cannot be filled with one number in each position if remainders on division by 6 are to be the same in each column and remainders on division by 4 are to be the same in each row.

22c Systematic catching practice – comments

When there are n cricketers around the circle for catching practice, throwing the ball around the circle in steps of l to the left (starting from position 0), the ball reaches the cricketers in positions 0, l, $l + l$, $l + l + l$, ... with n and multiples of n subtracted when possible.

In **17. Can you really tell the time?** you made an addition table *modulo* 12:

$$(Z_{12}, +)$$

+	0	1	2	3	4	5	6	7	8	9	10	11
0	0	1	2	3	4	5	6	7	8	9	10	11
1	1	2	3	4	5	6	7	8	9	10	11	0
2	2	3	4	5	6	7	8	9	10	11	0	1
3	3	4	5	6	7	8	9	10	11	0	1	2
4	4	5	6	7	8	9	10	11	0	1	2	3
5	5	6	7	8	9	10	11	0	1	2	3	4
6	6	7	8	9	10	11	0	1	2	3	4	5
7	7	8	9	10	11	0	1	2	3	4	5	6
8	8	9	10	11	0	1	2	3	4	5	6	7
9	9	10	11	0	1	2	3	4	5	6	7	8
10	10	11	0	1	2	3	4	5	6	7	8	9
11	11	0	1	2	3	4	5	6	7	8	9	10

From the table you can check which integers reach all 12 numbers by repeated addition. The ones that do $\{1, 5, 7, 11\}$ are called generators of addition *modulo* 12, or generators of $(Z_{12}, +)$. The ones that don't have a common factor with 12, greater than 1.

Either l and n have a common factor $d > 1$, and the sequence can only reach multiples of d, because if x and y are integers, $xl - yn$ always has a factor d;

or l and n have no common factor other than 1 (and are said to be **coprime**) so that from the Euclidean algorithm (in **6. Algorithma**) there are integers x and y such that $xl + yn = 1$, so $xl = 1 - yn$ and after x throws, the cricketer in position 1 gets the ball. After kx throws, the cricketer in position k gets the ball (though of course, it may have reached that position earlier).

If x were negative, note that $xl + yn = 1 \Rightarrow (n + x)l + (y - l)n = 1$, so that $n + x$ throws would reach 1.

$\phi(n)$ is a count of the whole numbers between 1 and n inclusive which are coprime to n (i.e. the quantity of different step lengths by which the ball reaches every position around the circle). Obviously $\phi(p) = p - 1$ for any prime number p, and in fact $\phi(p^n) = p^n - p^{n-1} = p^{n-1}(p - 1) = p^n(1 - 1/p)$.

Now $\phi(12) = \phi(3) \cdot \phi(4)$ but $\phi(12) \neq \phi(2) \cdot \phi(6)$. Also $\phi(18) = \phi(2) \cdot \phi(9)$ but $\phi(18) \neq \phi(3) \cdot \phi(6)$. For $\phi(12)$ we count

$$1, 2, 3, 4, \mathbf{5}, 6, 7, 8, 9, 10, \mathbf{11}, 12$$

which we can count more readily when displayed as

Remainders when divided by 4

		0	1	2	3
Remainders	0	12	9	6	3
when divided	1	4	1	10	7
by 3	2	8	5	2	11

The condition which allows us to claim that $\phi(p^n q^m) = \phi(p^n)\phi(q^m)$ for any two distinct prime numbers p and q is **the multiplicative property that** $\phi(m)\phi(n) = \phi(mn)$ **whenever m and n have no common factors**. This follows from the possibility of locating the mn numbers from 1 to mn in an $m \times n$ rectangular array with all the numbers in a column having the same remainder when divided by m and all the numbers in a row having the same remainder when divided by n. (See **21. Chinese remainders – comments**.) For then any number between 1 and mn having a common factor with mn must have a prime common factor, and a prime common factor is either a factor of m or a factor of n, but not both. Suppose that p is a prime factor of m (and therefore a prime factor of mn), and that $m = kp$. Then the columns of the rectangular $m \times n$ array of numbers with remainders $0, p, 2p, \ldots, (k-1)p$ on division by m contain all the numbers between 1 and mn with a factor p. If we wish to identify the numbers in the array which are coprime to mn we must delete the k columns containing the numbers with a factor p. As we then proceed with the other prime factors of mn, we must delete all but $\phi(m)$ columns of the array and all but $\phi(n)$ rows of the array, leaving just $\phi(m)\phi(n)$ numbers, so $\phi(m)\phi(n) = \phi(mn)$. Don't forget, you can only be sure of this when hcf(m, n) = 1.

The ϕ function was first introduced by L. Euler (1760) in his discussion of the ideas in **44. Non-zero products**, *though the use of the symbol ϕ for this function only dates from Gauss (1801)*.

23c Do you know your tables? – comments

$$(10k + a)(10l + b) = 100kl + 10kb + 10al + ab$$
$$= 10(10kl + kb + al) + ab$$

and $10(10kl + kb + la)$ does not affect the last digit.

$$x \equiv a \;(\mathrm{mod}\; 10) \Rightarrow x = 10k + a$$
$$y \equiv b \;(\mathrm{mod}\; 10) \Rightarrow y = 10l + b$$

So, $xy = (10k + a)(10l + b)$ and $xy = 10(10kl + kb + al) + ab \Rightarrow xy \equiv ab \;(\mathrm{mod}\; 10)$.

If cards numbered a and b were in the same player's hand (in **16. No shuffling**) then $a \equiv b \;(\mathrm{mod}\; 4)$.

$$x \equiv a \;(\mathrm{mod}\; 4) \Rightarrow x = 4k + a$$
$$y \equiv b \;(\mathrm{mod}\; 4) \Rightarrow y = 4l + b$$

So, $xy = (4k + a)(4l + b)$ and $xy = 4(4kl + kb + al) + ab \Rightarrow xy \equiv ab \;(\mathrm{mod}\; 4)$.

$$x \equiv a \;(\mathrm{mod}\; 3) \Rightarrow x = 3k + a$$
$$y \equiv b \;(\mathrm{mod}\; 3) \Rightarrow y = 3l + b$$

So, $xy = (3k + a)(3l + b)$ and $xy = 3(3kl + kb + al) + ab \Rightarrow xy \equiv ab \;(\mathrm{mod}\; 3)$.

modulo 3 $(\mathbf{Z}_3, \cdot)$

$\cdot$	0	1	2
0	0	0	0
1	0	1	2
2	0	2	1

$$x \equiv a \;(\mathrm{mod}\; n) \Rightarrow x = kn + a$$
$$y \equiv b \;(\mathrm{mod}\; n) \Rightarrow y = ln + b$$

So, $xy = (kn + a)(ln + b)$ and $xy = n(kln + kb + al) + ab \Rightarrow xy \equiv ab \;(\mathrm{mod}\; n)$. Thus $x \equiv a \;(\mathbf{mod}\; n)$ **and** $y \equiv b \;(\mathbf{mod}\; n) \Rightarrow xy \equiv ab \;(\mathbf{mod}\; n)$, and this is the justification for claiming that multiplication works consistently on columns of integers, arranged with congruent integers in the same column, and therefore there is such a thing as $(\mathbf{Z}_n, \cdot)$.

modulo 4 $(\mathbf{Z}_4, \cdot)$

$\cdot$	0	1	2	3
0	0	0	0	0
1	0	1	2	3
2	0	2	0	2
3	0	3	2	1

modulo 5 $(\mathbf{Z}_5, \cdot)$

$\cdot$	0	1	2	3	4
0	0	0	0	0	0
1	0	1	2	3	4
2	0	2	4	1	3
3	0	3	1	4	2
4	0	4	3	2	1

modulo 6 $(\mathbf{Z}_6, \cdot)$

$\cdot$	0	1	2	3	4	5
0	0	0	0	0	0	0
1	0	1	2	3	4	5
2	0	2	4	0	2	4
3	0	3	0	3	0	3
4	0	4	2	0	4	2
5	0	5	4	3	2	1

modulo 7 $(\mathbf{Z}_7, \cdot)$ modulo 8 $(\mathbf{Z}_8, \cdot)$

·	0	1	2	3	4	5	6
0	0	0	0	0	0	0	0
1	0	1	2	3	4	5	6
2	0	2	4	6	1	3	5
3	0	3	6	2	5	1	4
4	0	4	1	5	2	6	3
5	0	5	3	1	6	4	2
6	0	6	5	4	3	2	1

·	0	1	2	3	4	5	6	7
0	0	0	0	0	0	0	0	0
1	0	1	2	3	4	5	6	7
2	0	2	4	6	0	2	4	6
3	0	3	6	1	4	7	2	5
4	0	4	0	4	0	4	0	4
5	0	5	2	7	4	1	6	3
6	0	6	4	2	0	6	4	2
7	0	7	6	5	4	3	2	1

modulo 9 $(\mathbf{Z}_9, \cdot)$

·	0	1	2	3	4	5	6	7	8
0	0	0	0	0	0	0	0	0	0
1	0	1	2	3	4	5	6	7	8
2	0	2	4	6	8	1	3	5	7
3	0	3	6	0	3	6	0	3	6
4	0	4	8	3	7	2	6	1	5
5	0	5	1	6	2	7	3	8	4
6	0	6	3	0	6	3	0	6	3
7	0	7	5	3	1	8	6	4	2
8	0	8	7	6	5	4	3	2	1

If you look at the integers 1, 2, 3, 4, ... , n you can search

(i) for numbers which have inverses under multiplication;
(ii) for numbers a such that $a, 2a, 3a, \ldots \equiv 1, 2, 3, \ldots \pmod{n}$ in some order;
(iii) for numbers which generate $(\mathbf{Z}_n, +)$;
(iv) for numbers coprime to n.

mod 3: 1, 2
mod 4: 1, 3
mod 5: 1, 2, 3, 4
mod 6: 1, 5
mod 7: 1, 2, 3, 4, 5, 6

mod 8: 1, 3, 5, 7
mod 9: 1, 2, 4, 5, 7, 8
mod 10: 1, 3, 7, 9
mod 11: 1, 2, 3, 4, 5, 6, 7, 8, 9, 10

(iv) $\Rightarrow$ (i): If hcf(a, n) = 1, then there are integers x, y such that $ax + ny = 1$ and $ax \equiv 1$ (mod n) so a has an inverse, namely x.
(i) $\Rightarrow$ (iv): If $ax \equiv 1$ (mod n) then $ax = 1 + kn$, so $ax - kn = 1$ and hcf(a, n) = 1.
(i) $\Rightarrow$ (ii) for if $ax \equiv 1$, then $a(2x) \equiv 2$ and $a(3x) \equiv 3$ etc.
(ii) $\Rightarrow$ (i) for some x, $xa \equiv 1$, so a has an inverse.
(ii) $\Leftrightarrow$ (iii) for $2a = a + a$, $3a = a + a + a$, etc.

The congruence notation was devised by C. F. Gauss and the theorem we have stated in bold face above first appeared in section I of Gauss' Disquisitiones Arithmeticae (1801). This book is said to be the first mathematics text written with modern standards of rigour. It was published when Gauss was 24.

24c Coding and decoding – comments

To crack the code GPIZIV HMGO, we first try replacing each letter by the next one in the alphabet. To save time we translate only a couple of letters each time and only translate the whole word if the first two could form part of an English word. For example, IV at the end of the first word would become JW which is not a feasible word ending so we try the next: this gives KX again not feasible. You should have found that by moving the letters each forward by 22 gave the phrase "clever dick". Thus the encoder moved every letter forward 4. The next code is similar and this time the encoder moved every letter forward 3: the original message was "et tu Brute".

If we have a multiplier of 2 then the letters a and n, numbered 0 and 13 respectively, are both mapped to 0. Thus it is not possible to decode the message unambiguously. To ensure that we get a different letter for each of our initial 26 letters the multiplier p must be coprime to 26. From our work on Euler's ϕ function in **22. Systematic catching practice** we know that $\phi(26)$ is 12 and $p = 1, 3, 5, 7, 9, 11, 15, 17, 19, 21, 23$ or 25. Any number from 0 to 25 is suitable for the constant q. Thus there are $12 \times 26 = 312$ codes of this form.

In the code given, the letters H and Y occur the most frequently, A, N and Q occur twice and the rest once. Also the letter Q forms a word on its own. Using these facts and a little trial and error you will find the message:

this is a secret message!

25c Repacking – comments

The square trays of size between 1 and 50 are 1, 4, 9, 16, 25, 36 and 49 and these leave remainders 1, 4, 4, 1, 0, 1 and 4 on repackaging in cartons of 5. From our survey of small square trays there appear not to be any square numbers $x^2 \equiv 2$ or 3 (mod 5). We shall show that this is indeed the case in **26. Where have all the squares gone?** and **27. Where have all the squares come from?**.

The numbers 2 and 3 are called *non-squares modulo* 5. The numbers 0, 1 and 4 are called *squares modulo* 5.

If we repeat this *modulo* 7 the remainders are now 1, 4, 2, 2, 4, 1, 0 and so the *non-squares modulo* 7 appear to be 3, 5 and 6.

26c Where have all the squares gone? – comments

All products of the kind $(3n + a)(3m + b)$ for $a, b = 0, 1$ and 2 can be summarised in the following "multiplication table *modulo* 3":

$$(\mathbf{Z}_3, \cdot)$$

$\cdot$	0	1	2
0	0	0	0
1	0	1	2
2	0	2	1

The columns of the original display which contain squares can be found by inspecting the main diagonal of the table.

All products of the kind $(4n + a)(4m + b)$ for $a, b = 0, 1, 2$ and 3 can be summarised in the following "multiplication table *modulo* 4":

$$(\mathbf{Z}_4, \cdot)$$

$\cdot$	0	1	2	3
0	0	0	0	0
1	0	1	2	3
2	0	2	0	2
3	0	3	2	1

Again, the columns of the original display which contain squares can be found by inspecting the main diagonal of the table.

All products of the kind $(5n + a)(5m + b)$ for $a, b = 0, 1, 2, 3$ and 4 can be summarised in the following "multiplication table *modulo* 5":

$$(\mathbf{Z_5}, \cdot)$$

·	0	1	2	3	4
0	0	0	0	0	0
1	0	1	2	3	4
2	0	2	4	1	3
3	0	3	1	4	2
4	0	4	3	2	1

Again, the columns of the original display which contain squares can be found by inspecting the main diagonal of the table.

When the positive integers are displayed in n columns with first row $[0, 1, 2, \ldots, n-1]$, and a square number x^2 appears in the display in a column with a at its head, then a is called a *square* or *quadratic residue modulo n*, that is a is a possible remainder when a square number is divided by n.

27c Where have all the squares come from? – comments

With five columns, the squares in the column headed 1 come from the columns headed 1 and 4; the squares in the column headed 4 come from the columns headed 2 and 3.

	x	x^2
	0	0
	1	1
mod 5	2	4
	3	4
	4	1

	x	x^2
	0	0
	1	1
	2	4
mod 7	3	2
	4	2
	5	4
	6	1

With seven columns, the squares in the column headed 1 come from the columns headed 1 and 6; the squares in the column headed 2 come from the columns headed 3 and 4; the squares in the column headed 4 come from the columns headed 2 and 5.

In each case, the sum of the two headings equals the number of columns.

If x^2 and y^2 are in the same column, then they differ by a multiple of p. So $p \mid x^2 - y^2$. Thus $p \mid (x - y)(x + y)$, and since p is prime (using the lemma of **9. A return**

visit to Algorithma) either $p \mid x - y$ (in which case x and y are in the same column, or in fact equal if they are both column headings) or $p \mid x + y$, in which case $y = p - x$ (if they are both column headings). This can be summarised symbolically. If $x^2 \equiv y^2$ (mod p), then either $x \equiv y$ (mod p) or $p - x \equiv y$ (mod p).

The overall conclusion is that if we only look at the columns headed $1, 2, \ldots, p - 1$, pairs of columns (headed typically by x and $p - x$) have squares in the same column. So the numbers of columns with squares is exactly half the total (we are ignoring the 0 column). We are presuming that p is a prime number greater than 2; this makes p an odd number and so x is never equal to $p - x$.

28c How old is Grandma? – comments

We obtain the equation: $h^2 - g^2 = 119143$. Thus $g^2 = h^2 - 119143$. The answer to the problem depends on finding squares which are still squares after deduction of 119143.

Since g is positive, h must be at least the square root of 119143. Hence h is greater than 345.

As you can discover from the table not all numbers are candidates for being squares. The number 573 could not possibly be a square because it ends in a 3. The last digit of the product of two numbers comes from the multiplication of the last digits of the two numbers in the products. So in order to see what the last possible digits of a square can be we only need to consider the last digits of the squares of 1 to 10:

$$(10n + k)^2 \equiv k^2 \text{ (mod 10)}$$

This gives us 0, 1, 4, 5, 6 and 9 as possible last digits of squares.

The number $h^2 - 119143$ is also a square. The $h^2 - 3$ (mod 10) row in the table gives us the last digit of $h^2 - 119143$. We can thus rule out any values of h which do not result in a 0, 1, 4, 5, 6 or 9 in the last row. Hence possible last digits of h are 2, 3, 7 or 8. We try 347 as a first possible value for h and then 348, 352, $\ldots$. But $352^2 - 119143 = 69^2$ and so $h = 347$ and $g = 69$ is a possible solution.

Since $h^2 - g^2 = 119143$ we can see that $(h - g)(h + g) = 119143$ and so two factors of 119143 are 283 and 421 which are both prime.

To factorise 2279 we need to find h, g such that $h^2 - g^2 = 2279$. By drawing up a table as before but with the last row $h^2 - 9$ *modulo* 10 we can see that h can end in 2, 3, 5, 7, 8 or 0 and of course $h > \sqrt{2279}$. So $h > 47$. Try $h = 48$. $48^2 - 2279 = 25$. Thus $g = 5$ and $2279 = 53 \cdot 43$.

This method of factorisation will eventually work for any odd number. Let the number we wish to factorise be n and assume it can be written as the product rs. Then we can write $n = h^2 - g^2$ where $h = \frac{1}{2}(r + s)$ and $g = \frac{1}{2}(r - s)$. Since we start our search from $\sqrt{n}$ we shall find h more quickly if it has a factor close to $\sqrt{n}$.

To factorise 156379 we can decrease the possible values for h by looking at the last two digits of squares. As we have seen we only need to obtain the squares from 1 to 25. Thus the possible last two digits of squares are 00, 01, 04, 09, 16, 21, 24, 25, 29, 36, 41, 44, 49, 56, 61, 64, 69, 76, 81, 84, 89 or 96.

h	0	1	2	3	4	5	6	7	8	9	10	11	12	13	14	15
h^2	00	01	04	09	16	25	36	49	64	81	00	21	44	69	96	25
g^2	21	22	25	30	37	46	57	70	85	02	21	42	65	90	17	46

h	16	17	18	19	20	21	22	23	24	25
h^2	56	89	24	61	00	41	84	29	76	25
g^2	77	10	45	82	21	62	05	50	97	46

Hence the possible last two digits for h are 00, 50, 02, 48, 52, 98, 10, 40, 60, 90 and 20, 30, 70, 80. Since $\sqrt{156379} > 394$ the first few numbers to consider for h are 398, 400, … and in fact $h = 398$ and $g = 45$ giving $156379 = 443 \times 353$.

Fermat demonstrated his method by finding the factors of 2027651281.

29c Higher powers – comments

$$(5n)^2 = 5(5n^2)$$
$$(5n + 1)^2 = 5(5n^2 + 2n) + 1$$
$$(5n + 2)^2 = 5(5n^2 + 4n) + 4$$
$$(5n + 3)^2 = 5(5n^2 + 6n + 1) + 4$$
$$(5n + 4)^2 = 5(5n^2 + 8n + 3) + 1$$

So when the positive integers are arranged in five columns, squares only appear in the columns below 0, 1 and 4.

A fourth power is the square of a square. But squaring a number in the columns under 0, 1 or 4 gives a number in the columns under 0 or 1:

$(5n)^4 = 5(125n^4)$, $(5n + x)^4 = 5k + 1$, when $x = 1, 2, 3$ or 4
$(7n)^2 = 7k$; $(7n + 1)^2$ and $(7n + 6)^2 = 7k + 1$;
$(7n + 2)^2$ and $(7n + 5)^2 = 7k + 4$; $(7n + 3)^2$ and $(7n + 4)^2 = 7k + 2$
$(7n)^3 = 7k$; $(7n + 1)^3$, $(7n + 2)^3$ and $(7n + 4)^2 = 7k + 1$;
$(7n + 3)^3$, $(7n + 5)^3$ and $(7n + 6)^3 = 7k + 6$.

A sixth power is both a square and a cube, so $(7n + x)^6 = 7k + 1$, when $x = 1, 2, 3, 4, 5, 6$.

$$1^2 \equiv 12^2 \equiv 1 \ (\mathrm{mod}\ 13),\ 2^2 \equiv 11^2 \equiv 4 \ (\mathrm{mod}\ 13),\ 3^2 \equiv 10^2 \equiv 9 \ (\mathrm{mod}\ 13),$$
$$4^2 \equiv 9^2 \equiv 3 \ (\mathrm{mod}\ 13),\ \ 5^2 \equiv 8^2 \equiv 12 \ (\mathrm{mod}\ 13),\ 6^2 \equiv 7^2 \equiv 10 \ (\mathrm{mod}\ 13)$$

So $1^4 \equiv 12^4 \equiv 5^4 \equiv 8^4 \equiv 1 \ (\mathrm{mod}\ 13)$, $2^4 \equiv 11^4 \equiv 3^4 \equiv 10^4 \equiv 3 \ (\mathrm{mod}\ 13)$, $4^4 \equiv 9^4 \equiv 6^4 \equiv 7^4 \equiv 9 \ (\mathrm{mod}\ 13)$.

$$1^3 \equiv 3^3 \equiv 9^3 \equiv 1 \ (\mathrm{mod}\ 13), \qquad 2^3 \equiv 5^3 \equiv 6^3 \equiv 8 \ (\mathrm{mod}\ 13),$$
$$4^3 \equiv 10^3 \equiv 12^3 \equiv 12 \ (\mathrm{mod}\ 13),\ 7^3 \equiv 8^3 \equiv 11^3 \equiv 5 \ (\mathrm{mod}\ 13)$$

Twelfth powers are both cubes and fourth powers and $\{1, 3, 9\} \cap \{1, 5, 8, 12\} = \{1\}$, so for $x \equiv 1, 2, 3, \ldots, 12 \pmod{13}$, $x^{12} \equiv 1 \pmod{13}$.

The pattern which is emerging is that for a prime number p, $x^{p-1} \equiv 1 \pmod{p}$ when $x \equiv 1, 2, 3, \ldots, p-1 \pmod{p}$.

30c Just shuffling and then power (*modulo* 7) – comments

x	$=$	0	1	2	3	4	5	6
$0 \cdot x$	$\equiv$	0	0	0	0	0	0	0
$1 \cdot x$	$\equiv$	0	1	2	3	4	5	6
$2 \cdot x$	$\equiv$	0	2	4	6	1	3	5
$3 \cdot x$	$\equiv$	0	3	6	2	5	1	4
$4 \cdot x$	$\equiv$	0	4	1	5	2	6	3
$5 \cdot x$	$\equiv$	0	5	3	1	6	4	2
$6 \cdot x$	$\equiv$	0	6	5	4	3	2	1

The table has been exhibited previously in the notes on **section 23**. Each row, after the first row of products, starts with 0, but then contains all six numbers, from 1 to 6, in some order. We can describe the numbers in the $a \cdot x$ row. Of the seven numbers, $a \cdot 0$, $a \cdot 1$, $a \cdot 2$, $a \cdot 3$, $a \cdot 4$, $a \cdot 5$ and $a \cdot 6$, the first, $a \cdot 0 \equiv 0 \pmod 7$. But *if a does not have a factor 7*, then none of the six other integers has a factor 7, so each product is congruent to 1, 2, 3, 4, 5 or 6 *modulo* 7. We will prove that this must be the case.

If no two of these six numbers are congruent to each other, then each of $a \cdot 1$, $a \cdot 2$, $a \cdot 3$, $a \cdot 4$, $a \cdot 5$ and $a \cdot 6$ is congruent to a different one of 1, 2, 3, 4, 5 and 6, and multiplying by a shuffles the six numbers. So we check whether $a \cdot 1$, $a \cdot 2$, $a \cdot 3$, $a \cdot 4$, $a \cdot 5$ and $a \cdot 6$, are all different, *modulo* 7:

$$a \cdot x \equiv a \cdot y \pmod 7 \Rightarrow 7 \mid a \cdot x - a \cdot y$$
$$\Rightarrow 7 \mid a \cdot (x - y)$$
$$\Rightarrow 7 \mid a \text{ or } 7 \mid x - y$$

since 7 is a prime number. But we insisted that a must not have a factor 7, so $7 \mid x - y$ and $x \equiv y \pmod 7$. For the six numbers we were looking at, both x and y were between 1 and 6, so $\mid x - y \mid < 6$, and so $7 \mid x - y \Rightarrow x = y$. Provided a has no factor 7, $ax \equiv ay \pmod 7 \Rightarrow x \equiv y \pmod 7$, which is a *cancellation law*.

We have proved that each member of the set $\{a \cdot 1, a \cdot 2, a \cdot 3, a \cdot 4, a \cdot 5, a \cdot 6\}$ is congruent to a different member of the set $\{1, 2, 3, 4, 5, 6\}$ *modulo* 7. Thus $(a \cdot 1) \cdot (a \cdot 2) \cdot (a \cdot 3) \cdot (a \cdot 4) \cdot (a \cdot 5) \cdot (a \cdot 6) \equiv 1 \cdot 2 \cdot 3 \cdot 4 \cdot 5 \cdot 6 \pmod 7$, and so $(a^6) \cdot (1 \cdot 2 \cdot 3 \cdot 4 \cdot 5 \cdot 6) \equiv 1 \cdot 2 \cdot 3 \cdot 4 \cdot 5 \cdot 6 \pmod 7$. But the number $1 \cdot 2 \cdot 3 \cdot 4 \cdot 5 \cdot 6$ does not have a factor 7, so it may be cancelled and $a^6 \equiv 1 \pmod 7$.

31c Just shuffling and then ^{power} (*modulo* 3, 5, 11) – comments

x	$=$	1	2
$1 \cdot x$	$\equiv$	1	2
$2 \cdot x$	$\equiv$	2	1

modulo 3 $2^2 \equiv 1 \pmod 3$

x	$=$	1	2	3	4
$1 \cdot x$	$\equiv$	1	2	3	4
$2 \cdot x$	$\equiv$	2	4	1	3
$3 \cdot x$	$\equiv$	3	1	4	2
$4 \cdot x$	$\equiv$	4	3	2	1

modulo 5 $2^4 \equiv 3^4 \equiv 4^4 \equiv 1 \pmod 5$

x	$=$	1	2	3	4	5	6	7	8	9	10
$1 \cdot x$	$\equiv$	1	2	3	4	5	6	7	8	9	10
$2 \cdot x$	$\equiv$	2	4	6	8	10	1	3	5	7	9
$3 \cdot x$	$\equiv$	3	6	9	1	4	7	10	2	5	8
$4 \cdot x$	$\equiv$	4	8	1	5	9	2	6	10	3	7
$5 \cdot x$	$\equiv$	5	10	4	9	3	8	2	7	1	6
$6 \cdot x$	$\equiv$	6	1	7	2	8	3	9	4	10	5
$7 \cdot x$	$\equiv$	7	3	10	6	2	9	5	1	8	4
$8 \cdot x$	$\equiv$	8	5	2	10	7	4	1	9	6	3
$9 \cdot x$	$\equiv$	9	7	5	3	1	10	8	6	4	2
$10 \cdot x$	$\equiv$	10	9	8	7	6	5	4	3	2	1

modulo 11 $2^{10} \equiv 3^{10} \equiv 4^{10} \equiv 5^{10} \equiv 6^{10} \equiv 7^{10} \equiv 8^{10} \equiv 9^{10} \equiv 10^{10} \equiv 1 \pmod{11}$

For any prime number p, when a is an integer without a factor p, none of the $p-1$ integers $a \cdot 1$, $a \cdot 2$, $a \cdot 3$, ... , $a \cdot (p-1)$ are congruent to 0 or to any of the other numbers in the list, *modulo* p. The claim that none are $\equiv 0 \pmod p$ can be justified because $ax \equiv 0 \pmod{p} \Rightarrow p \mid ax \Rightarrow p \mid a$ or $p \mid x$, since p is prime. Both of these are false, since $1 \le x \le p - 1$, so ax is not $\equiv 0 \pmod{p}$.

We now show that no two of the numbers in the list are congruent to one another:

$$ax \equiv ay \pmod{p} \Rightarrow p \mid ax - ay$$
$$\Rightarrow p \mid a(x - y)$$
$$\Rightarrow p \mid a \text{ or } p \mid x - y$$

since p is prime. Now, $p \mid a$ is false, so $p \mid x - y$, and $x \equiv y \pmod{p}$. But $1 \le x, y \le p - 1$, so $x = y$. $ax \equiv ay \pmod{p} \Rightarrow x \equiv y \pmod{p}$ is a *cancellation law* for multiplication *modulo p*. Thus each of the $p - 1$ integers $a{\cdot}1$, $a{\cdot}2$, $a{\cdot}3$, $\ldots$, $a{\cdot}(p - 1)$ is congruent to a different integer in the list $1, 2, 3, \ldots, p - 1$, *modulo p*.

Thus $(a{\cdot}1){\cdot}(a{\cdot}2){\cdot}(a{\cdot}3){\cdot} \ldots {\cdot}(a{\cdot}(p - 1)) \equiv 1{\cdot}2{\cdot}3{\cdot} \ldots {\cdot}(p - 1) \pmod{p}$, just shuffling. So $a^{p-1}{\cdot}(1{\cdot}2{\cdot}3{\cdot} \ldots {\cdot}(p - 1)) \equiv 1{\cdot}2{\cdot}3{\cdot} \ldots {\cdot}(p - 1) \pmod{p}$, and since $1{\cdot}2{\cdot}3{\cdot} \ldots {\cdot}(p - 1)$ does not have a factor p, $a^{p-1} \equiv 1 \pmod{p}$, to the power $p - 1$.

That $a^p \equiv a \pmod{p}$ holds for all integers a, or that $a^{p-1} \equiv 1 \pmod{p}$ provided a does not have a factor p, was first claimed by Fermat in 1640 and is known as **Fermat's (little) theorem.**

The earliest published proof was given by Euler in 1739. Euler's proof used the binomial expansion as follows:

$$(a + 1)^p = a^p + pa^{p-1} + \frac{p(p - 1)}{2}a^{p-2} + \ldots + 1$$

The binomial coefficients are integers as in Pascal's triangle. For those coefficients whose numerators contain a factor p, *the denominator is* r! *for some* r < p. *So the factor* p *cannot be cancelled and we have*

$$(a + 1)^p - a^p - 1 \equiv 0 \pmod{p}$$

so

$$(a + 1)^p - (a + 1) \equiv a^p - a \pmod{p}$$

Since $1^p - 1 = 0$, *the result follows by induction.*

32c Factorials – comments

When p is an odd prime number and a is some fixed integer without a factor p, then since the $p - 1$ integers $a{\cdot}1$, $a{\cdot}2$, $a{\cdot}3$, $\ldots$, $a{\cdot}(p - 1)$ are each congruent to a different one of $1, 2, 3, \ldots, p - 1$ modulo p, there is exactly one number b from this second list such that $ab \equiv 1 \pmod{p}$. Since $ab = ba$, we have $ab = ba \equiv 1 \pmod{p}$.

The two numbers a and b are generally different. We find out when they can be equal. If $a = b$, then $a^2 \equiv 1 \pmod{p}$, so $p \mid a^2 - 1$, and $p \mid a - 1$ or $p \mid a + 1$. Thus either $a \equiv 1 \pmod{p}$ or $a \equiv -1 \equiv p - 1 \pmod{p}$.

So, except for $a = 1$ and $a = p − 1$, each $a = 1, 2, 3, \ldots$, or $(p − 1)$ is paired with a unique $b \neq a$ such that $ab \equiv 1 \pmod{p}$. This means that the integers 2, 3, $\ldots$, $(p − 2)$ split into pairs with products $\equiv 1 \pmod{p}$. So the product of all these numbers $\equiv 1 \pmod{p}$. Thus $1 \cdot 2 \cdot 3 \cdot \ldots \cdot (p − 2) \cdot (p − 1) \equiv p − 1 \equiv −1 \pmod{p}$.

In 1770 this theorem was stated by Waring to be due to Sir John Wilson and it is known as **Wilson's theorem.** *The first published proof was given by Lagrange in 1771.*

33c Square roots of −1, prime modulus – comments

There are no solutions of $x^2 + 1 \equiv 0$ *modulo* 3, 7, 11 or 19. However, $2^2 + 1 \equiv 0 \pmod{5}$, $5^2 + 1 \equiv 0 \pmod{13}$, $4^2 + 1 \equiv 0 \pmod{17}$.

The "no solution" primes are all of the form $4k + 3$, that is $\equiv 3 \pmod 4$. The primes (other than $p = 2$), with a solution, are all of the form $4k + 1$, that is $\equiv 1 \pmod 4$.

Let the prime $p = 4k + 3$, and suppose, if possible, that $x^2 + 1 \equiv 0 \pmod{p}$ for some integer x.

Then $x^2 \equiv −1 \pmod{p}$, so
$$x^4 \equiv 1 \pmod{p} \qquad\qquad (*)$$
But $x^{p−1} \equiv 1 \pmod{p}$ from Fermat's theorem.
So $x^{4k+2} \equiv 1 \pmod{p}$
and $(x^4)^k \cdot x^2 \equiv 1 \pmod{p}$,
so $1 \cdot (−1) \equiv 1 \pmod{p}$ from $(*)$, which is impossible since $p \neq 2$.

Let the prime $p = 4k + 1$;
then by Wilson's theorem $1 \cdot 2 \cdot 3 \cdot \ldots \cdot (p − 1) \equiv −1 \pmod{p}$.
So, $1 \cdot 2 \cdot 3 \cdot \ldots \cdot 2k \cdot (2k + 1) \cdot \ldots \cdot 4k \equiv −1 \pmod{p}$.
Now $2k + 1 = 4k + 1 − 2k$, so $2k + 1 \equiv −2k \pmod{p}$,
and $1 \cdot 2 \cdot 3 \cdot \ldots \cdot 2k \cdot −2k \cdot −(2k − 1) \cdot \ldots \cdot −1 \equiv −1 \pmod{p}$
But there are $2k$ negative integers on the left side of this congruence, an even number,
so $1 \cdot 2 \cdot 3 \cdot \ldots \cdot 2k \cdot 2k \cdot \ldots \cdot 3 \cdot 2 \cdot 1 \equiv −1 \pmod{p}$,
and $(1 \cdot 2 \cdot 3 \cdot \ldots \cdot 2k)^2 \equiv −1 \pmod{p}$.

For those who know the language of group theory, it may be worth pointing out that a square root of −1 has multiplicative order 4 and 4 is *not* a factor of $p − 1$ when $p = 4k + 3$, but 4 *is* a factor of $p − 1$ when $p = 4k + 1$.

34c How many square roots of −1? – comments

When a prime number $p = 4k + 3$, we have seen in **section 33** that there are no solutions of $x^2 + 1 \equiv 0 \pmod{p}$. There are no solutions of $x^2 + 1 \equiv 0 \pmod{15}$. A solution of $x^2 + 1 \equiv 0$ (mod $3n$) would have to be a solution of $x^2 + 1 \equiv 0$ (mod 3) and no such solution exists. If there is a solution of $x^2 + 1 \equiv 0$ (mod pn) where p is a prime number, $pn \mid x^2 + 1$, so $p \mid x^2 + 1$ and $p = 2$ or $p \equiv 1$ (mod 4), from **section 33**.

However, 2 and 3 are solutions of $x^2 + 1 \equiv 0$ (mod 5); 5 and 8 are solutions of $x^2 + 1 \equiv 0$ (mod 13); 4 and 13 are solutions of $x^2 + 1 \equiv 0$ (mod 17). 12 and 17 are solutions of $x^2 + 1 \equiv 0$ (mod 29).

When a prime $p = 4k + 1$, we have seen that there is at least one solution of $x^2 + 1 \equiv 0$ (mod p). But $(-x)^2 = x^2$, so if $x^2 + 1 \equiv 0$ (mod p), then $(p - x)^2 + 1 \equiv 0$ (mod p), and $x^2 + 1 \equiv 0$ (mod p) has at least two solutions.

Now, $x^2 + 1 = (x - \alpha)(x + \alpha) + \alpha^2 + 1$, and if $\alpha^2 + 1 \equiv 0$ (mod p), then $x^2 + 1 \equiv (x - \alpha)(x + \alpha)$ (mod p), which identifies two solutions.

If we suppose there is a third solution β, then $\beta^2 + 1 \equiv (\beta - \alpha) \cdot (\beta + \alpha)$ (mod p). Now $\beta^2 + 1 \equiv 0$, but

$$(\beta - \alpha) \cdot (\beta + \alpha) \equiv 0 \ (\text{mod } p) \Rightarrow p \mid (\beta - \alpha) \text{ or } p \mid (\beta + \alpha)$$
$$\Rightarrow \beta \equiv \alpha \text{ or } \beta \equiv -\alpha \ (\text{mod } p)$$

So our supposed third solution is one of those we have already found, and there are exactly two non-congruent solutions of $x^2 + 1 \equiv 0$ (mod p), for a prime $p = 4k + 1$.

$x = 2, 3$ give solutions of $x^2 + 1 \equiv 0$ (mod 5), and
$x = 5, 8$ give solutions of $x^2 + 1 \equiv 0$ (mod 13).

Using **21. Chinese remainders,**

$x \equiv 2$ (mod 5) and $x \equiv 5$ (mod 13) $\Rightarrow x \equiv 57$ (mod 65).
$x \equiv 2$ (mod 5) and $x \equiv 8$ (mod 13) $\Rightarrow x \equiv 47$ (mod 65).
$x \equiv 3$ (mod 5) and $x \equiv 5$ (mod 13) $\Rightarrow x \equiv 18$ (mod 65).
$x \equiv 3$ (mod 5) and $x \equiv 8$ (mod 13) $\Rightarrow x \equiv 8$ (mod 65).

$(x - 8)(x - 57) = x^2 - 65x + 456 \equiv x^2 + 1$ (mod 65), but neither $18 - 8$ nor $18 - 57$ has a factor 65. $18 - 8$ has a factor 5 and $18 - 57$ has a factor 13.

35c Sums of squares – comments

The square has area 13 which can be seen from the fact that its side has length $\sqrt{(10 - 8)^2 + (7 - 4)^2} = \sqrt{2^2 + 3^2}$. Thus the area of the square can be obtained from the coordinates of the two end points. We can express 4 as $2^2 + 0^2$ so this can be the area of a square with two vertices at (0,0) and (2,0); similarly $5 = 2^2 + 1^2$ and hence the vertices (0,0) and (2,1) will give a square with area 5. However, 6 cannot be expressed as the sum of two squares and so it is not possible to form a square of area 6 with corners on the grid points. To draw the square of area $2^2 + 2^2 = 8$ we can use the vertices (0,0) and (2,2).

+	0	1	4	9	16	25	36	49	64
0	0	*	*	*	*	*	*	*	*
1	1	2	*	*	*	*	*	*	*
4	4	5	8	*	*	*	*	*	*
9	9	10	13	17	*	*	*	*	*
16	16	17	20	25	32	*	*	*	*
25	25	26	29	34	41	50	*	*	*
36	36	37	40	45	52	61	72	*	*
49	49	50	53	58	65	74	85	98	*
64	64	65	68	73	80	89	100	113	128
81	81	82	85	90	97	106	117	130	145

With the integers in four columns the sums of squares all come in the columns headed 0, 1 or 2. The last column does not contain any sums of squares. The last column only contains numbers of the form $4k + 3$. All the squares come in the columns headed 0 or 1 because the square of an even number $(2n)^2 = 4n^2$ and the square of an odd number $(2n + 1)^2 = 4(n^2 + n) + 1$. So squares are always of the form $4k$ or $4k + 1$. By adding any two of these squares we can only obtain numbers of the form $4k$, $4k + 1$ or $4k + 2$. Hence the sum of two squares is never of the form $4k + 3$.

Hence we have proved that no numbers of the form $4k + 3$ are sums of squares and from our table it appears that primes of the form $4k + 1$ *are* sums of squares. Compare this result with that of **33. Square roots of –1, prime modulus**.

We have proved that an odd prime which is the sum of two squares is of the form $4k + 1$. Fermat also proved that every prime of the form $4k + 1$ could be expressed as the sum of two squares. How to prove this is shown in **53. Adding squares**.

36c   Sums of squares in two ways – comments

The numbers that appear twice in the table in the comments on **section 35** are 25, 50, 65 and 85. $25^2 = 3^2 + 4^2 = 5^2 + 0^2$ is not acceptable to the manufacturer as one of its sums uses 0. $50 = 7^2 + 1^2 = 5^2 + 5^2$ is not acceptable as one if its sums uses 5^2 twice.

$$65 = 5 \times 13$$
$$= (2^2 + 1^2)(3^2 + 2^2) = (2 \cdot 3 + 1 \cdot 2)^2 + (2 \cdot 2 - 1 \cdot 3)^2 = 8^2 + 1^2$$

and

$$= (1^2 + 2^2)(3^2 + 2^2) = (1 \cdot 3 + 2 \cdot 2)^2 + (1 \cdot 2 - 2 \cdot 3)^2 = 7^2 + 4^2$$

65 can be formed into two squares of 1×1 and 8×8 or two squares of 4×4 and 7×7.

$$85 = 5 \times 17$$
$$= (2^2 + 1^2)(4^2 + 1^2) = (2 \cdot 4 + 1 \cdot 1)^2 + (2 \cdot 1 - 1 \cdot 4)^2 = 9^2 + 2^2$$

and

$$= (1^2 + 2^2)(4^2 + 1^2) = (1 \cdot 4 + 2 \cdot 1)^2 + (1 \cdot 1 - 2 \cdot 4)^2 = 6^2 + 7^2$$

We can see that $65 = 13 \times 5$, $85 = 17 \times 5$ and $145 = 29 \times 5$ so each time the number is a product of two different primes both of the form $4k + 1$.

We have seen in **section 35** that no prime of the form $4k + 3$ is a sum of squares. An indication of the proof that all primes of the form $4k + 1$ are sums of squares is given in the comments on **section 53**.

We need to find a number which is the product of two sums of squares. The first such number greater than or equal to 150 that is the product of two different primes of the form $4k + 1$ is $185 = 37 \times 5$.

We can write

$$185 = 37 \times 5 = (2^2 + 1^2)(6^2 + 1^2) = (12 + 1)^2 + (6 - 2)^2 = 13^2 + 4^2$$
$$= (1^2 + 2^2)(6^2 + 1^2) = (6 + 2)^2 + (12 - 1)^2 = 8^2 + 11^2$$

37c Pythagorean triples – comments

Since $3^2 + 4^2 = 5^2$, $(3k)^2 + (4k)^2 = (5k)^2$, and $(3k\ 4k\ 5k)$ is a Pythagorean triple for any positive integer k.

If a, b, c are consecutive positive integers, $a = b - 1$ and $c = b + 1$, so if they form a Pythagorean triple, $(b - 1)^2 + b^2 = (b + 1)^2 \Rightarrow b^2 = 4b \Rightarrow b = 4$. So there is only one such triple.

If $a = Ad$ and $b = Bd$, then $c^2 = a^2 + b^2 = (Ad)^2 + (Bd)^2 = (A^2 + B^2)d^2$, so c has a factor d. Similarly if $a = Ad$ and $c = Cd$, then $b^2 = c^2 - a^2 = (Cd)^2 - (Ad)^2 = (C^2 - A^2)d^2$, so b has a factor d.

If $c = b + 1$, then $a^2 + b^2 = (b + 1)^2$, so $a^2 = 2b + 1$, and for any odd number a, such a b can be found. Look in the list at the places where $a = 3, 5, 7, 9$, etc., and locate consecutive numbers b and c.

If $a = b - 1$, $2b^2 - 2b + 1 = c^2$. This condition is satisfied comparatively rarely. The four smallest examples are $(3\ 4\ 5)$, $(20\ 21\ 29)$, $(119\ 120\ 169)$ and $(2089\ 2090\ 2955)$. It occurs when the ath triangular number is twice the $(\frac{1}{2}(c - 1))$th triangular number.

In the primitive Pythagorean triples listed, either a or b has a factor 3.

mod 3		b b^2	0 0	1 1	2 1
a	a^2	$+$			
0	0		0	1	1
1	1		1	2	2
2	1		1	2	2

But $c^2 \equiv 2$ (mod 3) is impossible, so either a or $b \equiv 0$ (mod 3) in every Pythagorean triple.

In the primitive Pythagorean triples listed, either a or b or c has a factor 5.

mod 5		b b^2	0 0	1 1	2 4	3 4	4 1
a	a^2	$+$					
0	0		0	1	4	4	1
1	1		1	2	0	0	2
2	4		4	0	3	3	0
3	4		4	0	3	3	0
4	1		1	2	0	0	2

But $c^2 \equiv 2$ and $c^2 \equiv 3$ (mod 5) are impossible, and for each of the remaining possibilities, either a or b or $c \equiv 0$ (mod 5). Thus every Pythagorean triple contains a number divisible by 5.

In the primitive Pythagorean triples listed, of a and b, one is odd and one is even. c is always odd. Either a or b has a factor 4.

mod 8		b b^2	0,4 0	1,3,5,7 1	2,6 4
a	a^2	$+$			
0,4	0		0	1	4
1,3,5,7	1		1	2	5
2,6	4		4	5	0

But $c^2 \equiv 2$ (mod 8) is impossible so a and b may not both be odd. $c^2 \equiv 5$ (mod 8) is impossible. $c^2 \equiv 0$ or 4 (mod 8) only occurs when the triple is not primitive. In the two remaining cases, either a or b has a factor 4 and the other is odd, and c is odd.

$2^2 + 1^2 = 5, 2^2 - 1^2 = 3, 2 \cdot 2 \cdot 1 = 4$

$3^2 + 2^2 = 13, 4^2 + 1^2 = 17, 4^2 + 3^2 = 25, 5^2 + 2^2 = 29, 6^2 + 1^2 = 37, 5^2 + 4^2 = 41,$

$7^2 + 2^2 = 53, 6^2 + 5^2 = 61, 8^2 + 1^2 = 7^2 + 4^2 = 65$

$$(p^2 - q^2)^2 + (2pq)^2 = p^4 - 2p^2q^2 + q^4 + 4p^2q^2$$
$$= p^4 + 2p^2q^2 + q^4.$$
$$= (p^2 + q^2)^2$$

So for any integers p and q, $(p^2 - q^2 \; 2pq \; p^2 + q^2)$ is a Pythagorean triple. $85 = 9^2 + 2^2$ $= 7^2 + 6^2$. So $(36\ 77\ 85)$ and $(13\ 84\ 85)$ are Pythagorean triples.

If $(a\ b\ c)$ is a primitive Pythagorean triple and a and c are odd, both $c + a$ and $c - a$ are even and $b^2 = (c + a)(c - a)$. Suppose $\text{hcf}(c + a, c - a) = 2d$. Then $2d \mid c + a$ and $2d \mid c - a$, so $d \mid (c + a) + (c - a) = 2c$ and $2d \mid (c + a) - (c - a) = 2a$, so $d \mid c$ and $d \mid a$. But $(a\ b\ c)$ is primitive, so $d \mid c$ and $d \mid a$ implies $d = 1$. So $\frac{1}{2}(c + a)$ and $\frac{1}{2}(c - a)$ are coprime, and since $(\frac{1}{2}b)^2 = [\frac{1}{2}(c + a)][\frac{1}{2}(c - a)]$, both $\frac{1}{2}(c + a)$ and $\frac{1}{2}(c - a)$ are squares. Let $p^2 = \frac{1}{2}(c + a)$ and $q^2 = \frac{1}{2}(c - a)$; then $a = p^2 - q^2$, $b = 2pq$ and $c = p^2 + q^2$.

In a primitive triple, c is odd, so one of p and q is odd and the other even. $(\text{even})^2 \equiv 0 \pmod 4$ and $(\text{odd})^2 \equiv 1 \pmod 4$, so $c \equiv 1 \pmod 4$. The primitive Pythagorean triples *modulo* 4 are $(0\ 1\ 1), (0\ 3\ 1), (1\ 0\ 1), (3\ 0\ 1)$.

A systematic method of constructing Pythagorean triples was known to the Babylonians (c. 1500 BC). The $(p^2 - q^2 \; 2pq \; p^2 + q^2)$ formula was known to Euclid (c. 300 BC) and Diophantus (c. AD 200) could prove that it was complete.

PART III

PART III

38c Squares and non-squares – comments

In the literature, *squares* are called *quadratic res.dues* and *non-squares* are called *quadratic non-residues*.

In the comments on **27. Where have all the squares come from?** we found that exactly half the numbers $1, 2, 3, \ldots, p - 1$ were *squares modulo p* and therefore exactly half were *non-squares*. So there are $\frac{1}{2}(p - 1)$ *squares* and $\frac{1}{2}(p - 1)$ *non-squares* in this list.

If a is a *square*, $a \equiv x^2$ (mod p), and if b is a square, $b \equiv y^2$ (mod p), so $ab \equiv x^2 y^2 \equiv (xy)^2$ (mod p), and the product of two squares is always a square.

If a is a *square*, then since each of the $p - 1$ numbers $a \cdot 1, a \cdot 2, a \cdot 3, \ldots, a \cdot (p - 1)$ is congruent to a different number from the list $1, 2, 3, \ldots, p - 1$, exactly half of these are *squares* and half are not. But there are $\frac{1}{2}(p - 1)$ *squares* in the list $1, 2, 3, \ldots, p - 1$, and a times each of these *squares* is again a *square*, giving the $\frac{1}{2}(p - 1)$ *squares* in the list $a \cdot 1, a \cdot 2, a \cdot 3, \ldots, a \cdot (p - 1)$. So all the remaining numbers are *non-squares*, and each product of the *square* a and a *non-square* must give a *non-square*.

Finally let b be a *non-square* and consider the $p - 1$ numbers $b \cdot 1, b \cdot 2, b \cdot 3, \ldots, b \cdot (p - 1)$. Each member of this list is congruent to a different member of the list $1, 2, 3, \ldots, p - 1$, and so, again, half are *squares* and half are *non-squares*. But we have just seen that a *non-square* times a *square* gives a *non-square*, and since there are $\frac{1}{2}(p - 1)$ of these, the remaining $\frac{1}{2}(p - 1)$ products, all *non-squares* times *non-squares*, give *squares*.

This proof is given in section IV of Disquisitiones Arithmeticae, *by C.F. Gauss (1801).*

39c Powers of squares and non-squares – comments

We will show that when p is an odd prime number, and $0 < x < p$, then

$$x^{\frac{1}{2}(p-1)} \equiv 1 \ (\text{mod } p) \text{ when } x \text{ is a square (congruence 1)}$$
$$\text{and } x^{\frac{1}{2}(p-1)} \equiv -1 \ (\text{mod } p) \text{ when } x \text{ is a non-square (congruence 2).}$$

Step 1 *Every such x satisfies one or other of these congruences.*

Fermat's theorem gives $x^{p-1} \equiv 1$ (mod p) for all these x. Notice that $p - 1$ is even. Now $x^{p-1} \equiv 1$ (mod p) $\Rightarrow p \mid x^{p-1} - 1 = (x^{\frac{1}{2}(p-1)} - 1)(x^{\frac{1}{2}(p-1)} + 1)$ so one or other congruence must follow for each x.

Step 2 *Every square x satisfies the first congruence.*

If $x \equiv y^2$, then $x^{\frac{1}{2}(p-1)} \equiv (y^2)^{\frac{1}{2}(p-1)} \equiv y^{p-1} \equiv 1$ (mod p). So every square x satisfies congruence 1. From **section 27** we know that there are $\frac{1}{2}(p - 1)$ non-congruent squares between 0 and p, so we have found $\frac{1}{2}(p-1)$ non-congruent solutions of $x^{\frac{1}{2}(p-1)} \equiv 1 \pmod{p}$.

Step 3 *Every non-square x satisfies the second congruence.*

This is harder to prove, but because of step 1 it will be enough if we can show that a non-square cannot satisfy the first congruence.

The method is to prove that

$$x^{\frac{1}{2}(p-1)} - 1 \equiv (x - sq1)(x - sq2)(x - sq3) \ldots (x - sq\tfrac{1}{2}(p-1)) \pmod{p}$$

where sq1, sq2, sq3, ... , sq$\frac{1}{2}(p-1)$ denote the $\frac{1}{2}(p-1)$ non-congruent squares *modulo p*. Then it will follow that putting $x = $ a non-square does not make any of the terms on the right hand side $\equiv 0 \pmod{p}$ and therefore cannot make their product (i.e. the left hand side) $\equiv 0 \pmod{p}$, so a non-square cannot satisfy the first congruence.

The proof is now complete apart from the claim that

$$x^{\frac{1}{2}(p-1)} - 1 \equiv (x - sq1)(x - sq2)(x - sq3) \ldots (x - sq\tfrac{1}{2}(p-1)) \pmod{p}$$

To establish this polynomial congruence we claim that integers $a, \ldots, k, l, m$ may be chosen to give the identity

$$\begin{aligned}
x^{\frac{1}{2}(p-1)} - 1 = {} & (x - sq1)(x - sq2)(x - sq3) \ldots (x - sq\tfrac{1}{2}(p-3))(x - sq\tfrac{1}{2}(p-1)) \\
& + a(x - sq1)(x - sq2)(x - sq3) \ldots (x - sq\tfrac{1}{2}(p-3)) \\
& + \ldots \\
& + k(x - sq1)(x - sq2) \\
& + l(x - sq1) \\
& + m
\end{aligned}$$

by considering the powers of x in turn.

Now because of step 2, putting $x = sq1$ shows that $m \equiv 0 \pmod{p}$. Now, putting $x = sq2$ shows that $l \equiv 0 \pmod{p}$, and so on. Finally, putting $x = sq\frac{1}{2}(p-1)$ shows that $a \equiv 0 \pmod{p}$, and the polynomial congruence is established.

The method of proof used in the last part of step 3 can be used in much more general situations to show that **a polynomial congruence of degree** n**, like** $x^n + ax^{n-1} + \ldots + lx + m \equiv 0 \pmod{p}$**, cannot have more than** n **non-congruent solutions, provided** p **is prime**.

The first proof of this result about polynomials of degree n was due to J.L. Lagrange in 1770. The method we have used is due to Crelle (1843).

40c The frequency of factors – comments

If n is a given positive integer and x must be chosen from the list 1, 2, ... , n, how many xs are there such that hcf$(n, x) = d$?

Firstly, unless d is a divisor of n, there are none at all. Secondly if hcf$(n, x) = d$, d is a divisor of n and of x, and hcf$(n/d, x/d) = 1$.

Now the number of integers between 1 and n/d which are coprime to n/d is $\phi(n/d)$, and $1 \le x/d \le n/d \Leftrightarrow d \le x \le n$. So $\phi(n/d)$ is the number of integers x with $1 \le x \le n$ such that hcf$(n, x) = d$.

We can count the numbers $1, 2, \ldots, n$ in two ways. Firstly, in the most obvious way, we get the result n. Secondly, we can count them, grouping the numbers according to their common factors with n. Then there are $\phi(n)$ with a common factor 1, and $\phi(n/d)$ with a common factor d. If we run through all the possible common factors, we will have included all the numbers from 1 to n just once.

If the factors of n are $1 = d_1, d_2, d_3, \ldots, d_m = n$, the numbers of integers with these common factors are $\phi(n/d_1), \phi(n/d_2), \phi(n/d_3), \ldots, \phi(n/d_m)$, respectively, and these are the numbers $\phi(d_1), \phi(d_2), \phi(d_3), \ldots, \phi(d_m)$ in some order, because $\{d_1, d_2, d_3, \ldots, d_m\}$ and $\{n/d_1, n/d_2, n/d_3, \ldots, n/d_m\}$ are the same set of integers. (If the first set of divisors is in increasing order, the second is in decreasing order.) Since $\phi(n/d_1) + \phi(n/d_2) + \phi(n/d_3) + \ldots + \phi(n/d_m) = n$, $\phi(d_1) + \phi(d_2) + \phi(d_3) + \ldots + \phi(d_m)$ $= n$. Isn't that a nice surprise?!

This result first appeared in section II of Disquisitiones Arithmeticae *published by C.F. Gauss in 1801, when he was 24 years old, and was believed to have been completed two years earlier by the "prince of mathematicians".*

41c Multiplication like addition – comments

We first prove that it is just integers x such that hcf$(n, x) = 1$ which are generators of the additive group *modulo n*.

Step 1 *If x is a generator, then hcf$(n, x) = 1$.*

Suppose the integers

$$x, x + x, x + x + x, \ldots, \underbrace{x + x + \ldots + x}_{n \text{ times}}$$

are congruent to each of the integers $0, 1, 2, \ldots, n - 1$ *modulo n*, in some order; then an integer congruent to 1 must appear in the first list.

So for some k, $kx \equiv 1 \pmod{n}$, and $n \mid kx - 1$. If $kx - 1 = ln$ for some l, and $kx - ln = 1$, then hcf$(n, x) = 1$.

Step 2 *If hcf$(n, x) = 1$, then x is a generator of the additive group modulo n.*

Here hcf$(n, x) = 1 \Rightarrow ax + bn = 1$ for some integers a and b. So $ax \equiv 1 \pmod{n}$, and an integer $\equiv 1$ appears in the first list, $ax \equiv 1 \pmod{n}$, so $(2a)x \equiv 2 \pmod{n}$, etc., and all the required numbers appear.

Thus x is a generator of $\{0, 1, 2, \ldots, n - 1\}$ for addition *modulo n*, precisely when hcf$(n, x) = 1$, and so there are just $\phi(n)$ generators in this group, $(\mathbf{Z}_n, +)$.

For multiplication *modulo n*, the story is not so tidy. Firstly, if *n* is not a prime, there may be numbers in the list $1, 2, \ldots, n-1$ for which repeated products sometimes give 0. Secondly, if *n* is a prime number *p*, the fact that we can find a generator *a* such that the integers

$$a, \; a \cdot a, \; a \cdot a \cdot a, \; \ldots, \; \underbrace{a \cdot a \cdot \ldots \cdot a}_{p-1 \text{ times}}$$

are congruent to the integers $1, 2, \ldots, p-1$, is something of a surprise, since there is no obvious reason why there should be one.

However, when a generator *a* has been found, the matching

$$1 \leftrightarrow a$$
$$2 \leftrightarrow a^2$$
$$3 \leftrightarrow a^3$$
$$\ldots$$
$$(\text{mod } p-1) \; 0 \equiv p-1 \leftrightarrow a^{p-1} \equiv 1 \;(\text{mod } p)$$

gives the kind of colour matching we have been able to exhibit between addition *modulo p − 1* and multiplication *modulo p*.

The generators for addition *modulo p − 1* are those *x* such that $\text{hcf}(p-1, x) = 1$ and the generators for multiplication *modulo p* are those a^x such that $\text{hcf}(p-1, x) = 1$, provided of course that a generator *a* has been found.

42c Powers to a prime modulus – comments

The numerical investigation suggested that under multiplication *modulo p* (a prime number), the order of a generator is $p-1$ and all other orders are divisors of this. We offer an algebraic proof.

Step 1 *The order of an integer not $\equiv 0 \;(\text{mod } p)$ is a factor of $p-1$.*

Each of the integers $1, 2, \ldots, p-1$ satisfies $x^{p-1} \equiv 1 \;(\text{mod } p)$, by Fermat's theorem, so the *order* of each of these integers is $p-1$ or less.

Suppose the integer *a* has order *k*, so that *k* is the least positive index such that $a^k \equiv 1 \;(\text{mod } p)$. Then $(a^k)^2 \equiv 1 \;(\text{mod } p)$, and more generally $a^{kn} \equiv 1 \;(\text{mod } p)$ for every integer *n*. Must $p-1$ be a multiple of *k*? Actually, yes, but for a proof, suppose not; then $p-1$ lies between two consecutive multiples of *k*, so $nk < p-1 < (n+1)k$, for some *n*, and then $a^{nk} \equiv 1 \equiv a^{p-1} \;(\text{mod } p)$. So $a^{(p-1)-nk} \equiv 1 \;(\text{mod } p)$. But $nk < p-1 < (n+1)k \Rightarrow 0 < (p-1) - nk < k$ so this index is less than *k*, contradicting the definition of *k* as the least index such that $a^k \equiv 1$. Consequently *k* must always be a factor of $p-1$.

Students who are already familiar with Lagrange's theorem in the theory of groups will not need the argument we have given here.

Thus, under multiplication *modulo p*, each of the integers $1, 2, 3, \ldots, p-1$ has order

$$either\ 1 = d_1, d_2, \ldots, or\ d_m = p - 1$$

where the d_i are the factors of $p - 1$.

Step 2 *There are either 0 or $\phi(d)$ non-congruent integers of order d.*

It may be that none of the numbers $1, 2, \ldots, p - 1$ has order d, but if there is one, let us call it a; then the congruence $x^d \equiv 1 \pmod{p}$ has at least one solution and, as in the comment on **39. Powers of squares and non-squares**, at most d solutions, because $x^d - 1$ is a polynomial of degree d.

But the powers of a, namely $a, a^2, a^3, \ldots, a^d \equiv 1$, are all distinct solutions of this congruence; therefore they are the d solutions (the most there can be) and thus include all numbers of order d. But, using the ideas of **section 41**, these powers of a multiply together *modulo p* like $1, 2, \ldots, d$ add together *modulo d*. Now, as in the conclusion of the comment on **section 41**, in the *additive* group (mod d) i is a generator when $hcf(d, i) = 1$, so in the multiplicative group a^i is a generator when $hcf(d, i) = 1$ and there are just $\phi(d)$ such elements. The generators of $\{a, a^2, a^3, \ldots, a^d \equiv 1\}$ are the numbers of order d. So if amongst the integers $1, 2, \ldots, p - 1$ there is one element of order d, there are exactly $\phi(d)$ elements of order d.

So for each possible order d, that is for each factor of $p - 1$, there are either no elements of that order, or else $\phi(d)$ elements of that order.

List of factors of $p - 1$: $1 = d_1, d_2, \ldots, d_m = p - 1$
Number of elements
of this order: 0 or $\phi(d_1)$, 0 or $\phi(d_2)$, ..., 0 or $\phi(d_m)$

But in **40. The frequency of factors** we found that

$$\phi(d_1) + \phi(d_2) + \phi(d_3) + \ldots + \phi(d_m) = p - 1$$

and every one of the integers $1, 2, \ldots, p - 1$ has an order d, which by step 1 is a divisor of $p - 1$. We will only have counted them all if there is an element of each possible order, and in fact the maximum number of elements of each order exist. This means there has to be an element of order $p - 1$, and this element is a *generator* of the multiplicative group *modulo p*.

A generator of this group is also called a *primitive root modulo p*. It is a solution of $x^{p-1} \equiv 1 \pmod{p}$, but not a solution of $x^n \equiv 1 \pmod{p}$ when $0 < n < p - 1$.

When $p = 4k + 1$, 4 is a divisor of $p - 1$ and so an element of *order* 4 must exist, which is what we need if -1 is to be a square.

The proof that there is a primitive root to a prime modulus first appeared in section III of Disquisitiones Arithmeticae *by C.F. Gauss (1801).*

43c Zero products – comments

The numerical exploration suggests

1. When the modulus is composite, zero products occur with neither component of the product $\equiv 0$.

2. When hcf(n, a) > 1, the list $a·1$, $a·2$, $a·3$, . . . , $a·(n - 1)$ contains at least one integer ≡ 0 (mod n).
3. A zero product $a·b$ ≡ 0 (mod n) can only occur (with $0 < a < n$ and $0 < b < n$) when both hcf(n, a) > 1 and hcf(n, b) > 1.

Explanation

1. When n is composite, say $n = ab$, then ab ≡ 0 (mod n).
2. If hcf(n, a) = d > 1 and $0 < a < n$, then $a(n/d) = (a/d)n$ ≡ 0 (mod n).
3. If ab ≡ 0 (mod n) and $0 < a$, $b < n$, then let us suppose (if possible) that hcf(n, a) = 1. Then there exist integers x and y such that $xn + ya = 1$, and so $xnb + yab$ = b. But both the terms on the left hand side have a factor n, so the right hand side must have a factor n. Thus b ≡ 0 (mod n) and this contradicts $0 < b < n$. So the supposition is wrong and hcf(n, a) > 1, Likewise hcf(n, b) > 1.

44c Non-zero products – comments

We can avoid any possibility of zero products *modulo n*, if we avoid integers a such that hcf(n, a) > 1. So we make a list of the integers from 1 to n which are coprime to n, and call it M:

$$M = \{a \mid 1 \le a \le n, \text{hcf}(n, a) = 1\}$$

There are $\phi(n)$ integers in M. We show that multiplying all the numbers in M by one of them shuffles the numbers in M *modulo n*.

Step 1 *If* a ∈ M *and* b ∈ M, *then* ab ≡ *some element of* M *(mod n).*

Hcf(n, a) = 1 and hcf(n, b) = 1 implies hcf(n, ab) = 1 (for if ab and n had a common factor, they would have a common prime factor p, say, and then $p \mid ab$ implies $p \mid a$ or $p \mid b$ which would contradict either hcf(n, a) = 1 or hcf(n, b) = 1).

Step 2 *If* a, b, c *are in* M *then* ab ≡ ac *(mod n) implies* b ≡ c *(mod n).*

If ab ≡ ac (mod n, then $n \mid ab - ac = a(b - c)$. But, hcf($n$, a) = 1 implies $n \mid b - c$ (from the comments on **section 9**) and so b ≡ c (mod n). So we have a cancellation law for multiplication in M.

So if b_1, b_2, b_3, . . . , b_m are the elements of M and a ∈ M, then $a·b_1$, $a·b_2$, $a·b_3$, . . . , $a·b_m$, are all congruent (mod n) to elements of M, from step 1; and $a·b_1$, $a·b_2$, $a·b_3$, . . . , $a·b_m$ are all congruent to different elements of M, from step 2.

$$\text{So } (a·b_1)·(a·b_2)·(a·b_3)· \ldots ·(a·b_m) \equiv b_1·b_2·b_3· \ldots ·b_m \ (mod \ n),$$
$$\text{and } (a^m)·(b_1·b_2·b_3· \ldots ·b_m) \equiv b_1·b_2·b_3· \ldots ·b_m \ (mod \ n).$$

Now we can use the cancellation law again to get a^m ≡ 1 (mod n).
Overall we have shown that when hcf(n, a) = 1, then $a^{\phi(n)}$ ≡ 1 (mod n).

This generalisation of Fermat's theorem is due to Euler (1761) and is known as the **Fermat–Euler theorem.**

45c Decimals to the death – comments

Terminators

$1/2 = 0.5$, $3/5 = 0.6$, $43/20 = 2.15$, $7/25 = 0.28$, $51/32 = 1.59375$, $137/50 = 2.74$, $1003/400 = 2.5075$, $1003/500 = 2.006$.

$3.5 = 7/2$, $4.7 = 47/10$, $0.12 = 3/25$, $2.45 = 49/20$, $0.375 = 3/8$, $1.054 = 527/500$, $30.3125 = 485/16$, $0.16 = 4/25$.

Every terminating decimal equals $N/2^a \cdot 5^b$, in lowest terms, for some integer N, where a and b are positive integers or 0.

There are two things to prove:

(i)   every terminating decimal is equal to a fraction of the form $N/2^a \cdot 5^b$;
(ii) every fraction of the form $N/2^a \cdot 5^b$ is equal to a terminating decimal.

(i)   Suppose a terminating decimal t has c digits after the decimal place; then $t \cdot 10^c$ is an integer M (say).

Thus $t = M/10^c = M/2^c \cdot 5^c$, and only 2s and 5s may be cancelled as there are no other primes in the denominator. Thus for some $a \le c$ and $b \le c$, $t = N/2^a \cdot 5^b$.

(ii) To show that a fraction $N/2^a \cdot 5^b$ equals a terminating decimal, it will be sufficient to show that $N/2^a \cdot 5^b$ equals an integer divided by a power of 10.

Suppose $a \ge b$; then

$$\frac{N}{2^a \cdot 5^b} = \frac{N \cdot 5^{a-b}}{2^a \cdot 5^b \cdot 5^{a-b}}$$

$$= \frac{N \cdot 5^{a-b}}{10^a} = \frac{M}{10^a} = \text{a terminating decimal}$$

A similar argument holds when $a < b$.

Cell block 1

$1/3 = 0.\dot{3}$, $2/3 = 0.\dot{6}$, $1/6 = 0.1\dot{6}$, $5/6 = 0.8\dot{3}$, $2/9 = 0.\dot{2}$, $8/9 = 0.\dot{8}$, $17/12 = 1.41\dot{6}$, $17/15 = 1.1\dot{3}$, $7/18 = 0.3\dot{8}$, $29/24 = 1.2083\dot{3}$.

Let $x = 1.333\dot{3}$; then $10x = 13.333\dot{3}$, so $10x - x = 12$, and $x = 4/3$, $0.0\dot{3} = 1/30$, $0.00\dot{3} = 1/300$, $1.1\dot{4} = 103/90$, $2.2\dot{3} = 67/30$, $0.\dot{1} = 1/9$, $0.1234\dot{5} = 11111/90000$.

A decimal with one recurring digit equals $N/2^a \cdot 5^b \cdot 3^c$, in lowest terms, for some integer N, where $c = 1$ or 2, and a and b are any positive integers or 0.

$10x - x = \text{terminating decimal} = N/2^a \cdot 5^b$. So $x = N/2^a \cdot 5^b \cdot 9$. Possibly one 3 may be cancelled, but not two 3s or else the decimal for x would terminate.

$0.abc999\dot{9} = 0.ab(c + 1) = \text{terminating decimal}$. The recurring nines must be preceded either by the decimal point, or by a digit $c < 9$. If $x = 0.abc999\dot{9}$, then $10x = a.bc999\dot{9}$, so

$$10x - x = a + \frac{b-a}{10} + \frac{c-b}{10^2} + \frac{9-c}{10^3}$$

$$9x = a \cdot \frac{9}{10} + \frac{b}{10} \cdot \frac{9}{10} + \frac{c}{10^2} \cdot \frac{9}{10} + \frac{9}{10^3}$$

$$x = a\left(\frac{1}{10}\right) + b\left(\frac{1}{10^2}\right) + c\left(\frac{1}{10^3}\right) + \frac{1}{10^3} = 0.ab(c+1)$$

Cell block 2

$2/11 = 0.\dot{1}\dot{8}$, $3/22 = 0.1\dot{3}\dot{6}$, $4/33 = 0.\dot{1}\dot{2}$, $5/44 = 0.11\dot{3}\dot{6}$, $6/55 = 0.1\dot{0}\dot{9}$, $7/66 = 0.10\dot{6}$, $9/88 = 0.102\dot{2}\dot{7}$, $10/99 = 0.\dot{1}\dot{0}$, $101/176 = 0.5738\dot{6}\dot{3}$.

$0.\dot{0}\dot{9} = 1/11$, $1.\dot{1}\dot{2} = 37/33$, $1.8\dot{1}\dot{2} = 299/165$, $1.87\dot{1}\dot{2} = 247/132$, $1.0\dot{2}\dot{7} = 113/110$, $2.12\dot{1}\dot{6} = 5251/2475$.

A decimal with a pair of recurring digits equals $N/2^a \cdot 5^b \cdot 3^c \cdot 11$, in lowest terms, for some integer N, where $c = 0$, 1 or 2 and a and b are any positive integers or 0. $100x - x =$ terminating decimal $= N/2^a \cdot 5^b$. So $x = N/2^a \cdot 5^b \cdot 99$. Possibly one or two 3s may be cancelled, but not 11 or else x would have less than two recurring digits.

Cell block 6

$1/7 = 0.\dot{1}4285\dot{7}$, $2/7 = 0.\dot{2}8571\dot{4}$, $3/7 = 0.\dot{4}2857\dot{1}$, $1/13 = 0.\dot{0}7692\dot{3}$, $2/13 = 0.\dot{1}5384\dot{6}$, $3/14 = 0.2\dot{1}4285\dot{7}$, $5/21 = 0.2\dot{3}8095\dot{...}$, $3/26 = 0.1\dot{1}5384\dot{6}$, $13/28 = 0.46\dot{4}2857\dot{1}$, $8/77 = 0.\dot{1}0389\dot{6}$.

$999999 = 3 \cdot 3 \cdot 3 \cdot 7 \cdot 11 \cdot 13 \cdot 37$ (but $999 = 3 \cdot 3 \cdot 3 \cdot 37$). $0.\dot{7}1428\dot{5} = 5/7$, $1.\dot{2}3076\dot{9} = 16/13$, $0.3\dot{5}7142\dot{8} = 5/14$, $0.3\dot{8}0952 = 8/21$, $0.1\dot{9}2307\dot{6} = 5/26$, $0.\dot{1}2820\dot{5} = 5/39$.

Decimals with a block of six recurring digits equal $N/2^a \cdot 5^b \cdot 3^c \cdot 11^d \cdot 37^e \cdot 7^f \cdot 13^g$, in lowest terms, where a and b are any positive integers or 0; $c = 0$, 1, 2 or 3; d, e, f and $g = 0$ or 1; and either f or g or both $= 1$, or $d = 1$ and $c = 3$, or $e = 1$ or both. $1000000x - x =$ terminating decimal $= N/2^a \cdot 5^b$. So $x = N/2^a \cdot 5^b \cdot 999999$.

If x is a decimal with a block of n recurring digits, then $10^n x - x =$ a terminating decimal, and this shows that every recurring decimal has an equivalent fraction.

$$\frac{T}{MY} = \frac{REPEAT}{999999}$$

What can MY be? It must be an odd number with a factor of 7 or 13 or both (for if not, it would have both 11 and 27 or 37 as factors, and that would make it too big). The only multiples of 7 that are worth trying are 21, 63 and 91. The only multiples of 13 that are worth trying are 13, 39 and 91. Then look at the last digit of $(999999/MY) \cdot T$. Use a calculator and start running through the possibilities for T: single digits with no factor in common with MY. Very few of the possibilities for MY give the pattern of Es in $REPEAT$.

When you have completed this section you should have learnt that you can convert a recurring decimal with a recurring block of length n into a fraction by multiplying by 10^n and subtracting the original decimal.

Recurring decimals were discussed by John Wallis in 1676 and by Leibniz in 1677. J.H. Lambert (1758) showed that every recurring decimal is equal to a rational fraction.

46c Recurring decimals – comments

$$\frac{12345}{99999} = 0.\dot{1}234\dot{5}, \quad \frac{1234}{99999} = 0.\dot{0}123\dot{4}, \quad \frac{123}{99999} = 0.\dot{0}012\dot{3},$$

$$\frac{521}{11111} = \frac{4689}{99999} = 0.\dot{0}468\dot{9}, \quad \frac{1}{9.41} = \frac{271}{99999} = 0.\dot{0}027\dot{1}$$

$$\frac{1}{41} = \frac{9.271}{99999} = \frac{2439}{99999} = 0.\dot{0}243\dot{9}, \quad \frac{1}{410} = \left(\frac{1}{10}\right)\left(\frac{2439}{99999}\right) = 0.00\dot{2}43\dot{9}$$

$$\frac{1}{82} = \frac{5}{410} = \left(\frac{1}{10}\right)\left(\frac{5.2439}{99999}\right) = \left(\frac{1}{10}\right)\left(\frac{12195}{99999}\right) = 0.0\dot{1}219\dot{5}$$

$$\frac{1}{164} = \frac{25}{4100} = \left(\frac{1}{100}\right)\left(\frac{25.2439}{99999}\right) = \left(\frac{1}{100}\right)\left(\frac{60975}{99999}\right) = 0.00\dot{6}097\dot{5}$$

$$\frac{1}{205} = \frac{2}{410} = \left(\frac{1}{10}\right)\left(\frac{2.2439}{99999}\right) = \left(\frac{1}{10}\right)\left(\frac{4878}{99999}\right) = 0.00\dot{4}87\dot{8}$$

$$\frac{1}{1025} = \frac{4}{4100} = \left(\frac{1}{100}\right)\left(\frac{4.2439}{99999}\right) = \left(\frac{1}{100}\right)\left(\frac{9756}{99999}\right) = 0.000\dot{9}75\dot{6}$$

Dealing with nine and ten digit numbers on a calculator may not be feasible. However, 999,999,999 = 999·1001001 and 9,999,999,999 = 99,999·100001. 9,999,999,999/9091 = 1099989, so 1/9091 = 0.\dot{0}00109998\dot{9}. 2/9091, 1/2·9091, 3/9091, 1/3·9091, 1/1099989 each have a recurring block of ten digits.
    1/17 = N/(sixteen digits all nines), so as a recurring decimal 1/17 has a recurring block of sixteen digits. $10^{16} \equiv 1$ (mod 17) by Fermat's theorem (**31. Just shuffling and then** [power]).

The decimal for $1/n$ is terminating when $n = 1, 2, 4, 5, 8, 10, 16, 20, 25, 32, 40, 50$.
Single recurring digit when $n = 3, 6, 9, 12, 15, 18, 24, 30, 36, 45, 48$.
Block of two recurring digits when $n = 11, 22, 33, 44$.
Block of three recurring digits when $n = 27, 37$.
Block of five recurring digits when $n = 41$.
Block of six recurring digits when $n = 7, 13, 14, 21, 26, 28, 35, 39, 42$.

A pocket calculator will not enable you to find the length of recurring blocks longer than 8 with any ease.

Block of 15 recurring digits when $n = 31$.
Block of 16 recurring digits when $n = 17, 34$.
Block of 18 recurring digits when $n = 19, 38$.
Block of 21 recurring digits when $n = 43$.
Block of 22 recurring digits when $n = 23, 46$.
Block of 28 recurring digits when $n = 29$.
Block of 42 recurring digits when $n = 49$.
Block of 46 recurring digits when $n = 47$.

$1/n$ and $2/n$ always have recurring blocks of the same length. This follows from the results of **45. Decimals to the death**.

$1/n$ and $3/n$ usually have recurring blocks of the same length, but not if the prime factorisation of n contains 3 to a positive index not equal to 2. For recurring blocks of length less than or equal to 6, this follows from the results of **45. Decimals to the death**.

1, 2	1, 2	6, 6
1, 6	2, 10	1, 2
6, 12	6, 6	1, 2
16, 16	1, 6	6, 12
2, 10	1, 2	6, 12
3, 18	6, 6	1, 2

In shifting from R to ϕ, powers of 2 and 5 have been removed because these factors do not affect the length of the recurring block. Apparently $R(n)$ is a factor of $\phi(m)$ when $n = m \cdot 2^a \cdot 5^b$ and neither 2 nor 5 is a factor of m.

$10^n - 1$ has n nines in a row. $10^n \equiv 1 \pmod{m}$. $10^{p-1} \equiv 1 \pmod{p}$ by Fermat's theorem. Thus an integer consisting of $p - 1$ digits all equal to $9 = pk$ for some k.

$1/p = k/(p-1 \text{ nines}) = 0.(a$ *recurring block of $p - 1$ digits ending with the digits of k*)

In some cases we may also have $10^n \equiv 1 \pmod{p}$ when n is a factor of $p - 1$. Thus if $p = 13$, $n = 6$ satisfies the congruence.

$$\frac{7}{550} = \frac{7}{50 \cdot 11} = \frac{14}{100 \cdot 11} = \left(\frac{1}{100}\right)\left(1 + \frac{3}{11}\right) = \left(\frac{1}{100}\right)\left(1 + \frac{27}{99}\right)$$

$$= \left(\frac{1}{100}\right)(1.\dot{2}\dot{7}) = 0.01\dot{2}\dot{7}$$

If $M = 2^a \cdot 5^b \cdot K$ with hcf$(10, K) = 1$ and $a \geq b$, then

$$\frac{N}{M} = \frac{5^{a-b}N}{10^a K} = \frac{L}{10^a K}$$

(There is a similar proof when $b > a$.)

By the Fermat–Euler theorem (in **44. Non-zero products**) we have $10^{\phi(K)} \equiv 1 \pmod{K}$. This means that K is a factor of $10^{\phi(K)} - 1$. Let $KJ = 10^{\phi(K)} - 1$ be an integer consisting of $\phi(K)$ nines. Now $L/K = \text{integer} + r/K$, where $0 \leq r < K$, and

L/K = integer + rJ/KJ. Now rJ/KJ is a decimal with a recurring block of digits of length $\phi(K)$, so the same is true of L/K, and hence of N/M, by shifting the decimal point a places.

When these ideas are combined they enable us to show that every rational number may be expressed as a terminating or as a recurring decimal.

Every integer which does not have a factor of 2 or a factor of 5 is a divisor of a number in the sequence $9, 99, 999, 9999, 99999, \ldots$.

John Wallis noted in 1676 that the rational fraction M/N gives rise to a recurring decimal with a recurring block of length at most $N - 1$ digits.

47c Can you reveal all and keep it secret? – comments

The completed table shows that each number is mapped to a unique number under the operation cube *modulo* 11 and that the operation seventh power *modulo* 11 gives the inverse mapping.

x	0	1	2	3	4	5	6	7	8	9
$x^3 \pmod{11}$	0	1	8	5	9	4	7	2	6	3
$(x^3)^7 \pmod{11}$	0	1	2	3	4	5	6	7	8	9

The message 5031 encodes to 4051.

Not all powers give unique numbers for each digit as the table shows if we take the power 2:

x	0	1	2	3	4	5	6	7	8	9
$x^2 \pmod{11}$	0	1	4	9	5	3	3	5	9	4

In arithmetic *modulo* 11 the powers that give unique digits are 1, 3, 7 and 9, that is to say the numbers coprime to $10 = \phi(11)$.

$$(x^3)^7 = x^{21} = x^{10}x^{10}x \equiv x \pmod{11}$$

using Fermat's theorem.

If we work *modulo* 13 then $\phi(13)$ is 12 and if we encode using the power 5 we need the power 5 to decode since $5 \cdot 5 = 25 = 2 \cdot 12 + 1$.

If $m = 35$ and $a = 7$ then $\phi(35) = \phi(7)\phi(5) = 24$ and $7 \cdot 7 = 2 \cdot 24 + 1$, so $b = 7$.

Once we know that $2027651281 = 44021 \cdot 46061$ and that these are both primes we can work out $\phi(2027651281) = \phi(44021)\phi(46061) = 44020 \cdot 46060$. So $3b \equiv 1 \pmod{2027561200}$ and hence the decoder b is 1351707467.

The message COME HOME would be initially written as the numbers 0315 1305 0815 1305 and then each four digit number would be encoded.

48c Primes as squares and non-squares – comments

		Squares or non-squares						
		3	5	7	11	13	17	19
	3	*	N	**Y**	**N**	Y	N	**Y**
	5	N	*	N	Y	N	N	Y
	7	**N**	N	*	**Y**	N	N	**N**
modulus	11	**Y**	Y	**N**	*	N	N	**N**
	13	Y	N	N	N	*	Y	N
	17	N	N	N	N	Y	*	Y
	19	**N**	Y	**Y**	**Y**	N	Y	*

The bold (representing circled) entries are in the rows and columns labelled 3, 7, 11 and 19. Notice that $3 \equiv 7 \equiv 11 \equiv 19 \equiv 3 \pmod 4$ and $5 \equiv 13 \equiv 17 \equiv 1 \pmod 4$.

Our conjecture is that when $p \equiv q \equiv 3 \pmod 4$ then p is a square *modulo q* if and only if q is a non-square *modulo p* but that when either p or $q \equiv 1 \pmod 4$ then p is a square *modulo q* if and only if q is a square *modulo p*.

We can introduce some notation to make this clearer. Let $(p \mid q) = 1$ when p is a square *modulo q* and let $(p \mid q) = -1$ when p is a non-square *modulo q*. The symbol $(p \mid q)$ is called the *Legendre* symbol.

We saw that $(p \mid q) \equiv p^{\frac{1}{2}(q-1)} \pmod q$ in **39. Powers of squares and non-squares**. So our conjecture can be stated as

$(p \mid q) = (q \mid p)$ when either p or q is congruent to 1 (mod 4).
$(p \mid q) = -(q \mid p)$ when both p and q are congruent to 3 (mod 4).

This is known as **Gauss' law of quadratic reciprocity**.

In fact we can define $(a \mid q)$ for any integer a and prime q. $(a \mid q) = 0$ if q divides a. Otherwise $(a \mid q) = 1$ when a is a square *modulo q* and $(a \mid q) = -1$ when a is not a square *modulo q*. These definitions are summarised in the congruence

$$(a \mid q) \equiv a^{\frac{1}{2}(q-1)} \pmod q$$

So $(ab \mid q) = (a \mid q)(b \mid q)$ which was also established in **38. Squares and non-squares**. In particular $(-1 \mid p) = 1$ when $p \equiv 1 \pmod 4$ and $(-1 \mid p) = -1$ when $p \equiv 3 \pmod 4$ from **section 33**.

The completed table

x	2	3	4	5	6	7	8	9	10
x^2	4	9	16	25	36	49	64	81	100
$x^2 - 2$	2	7	14	23	34	47	62	79	98
prime factors	2	7	2,7	23	2,17	47	2,31	79	2,7

x	11	12	13	14	15	16	17	18	19	20
x^2	121	144	169	196	225	256	289	324	361	400
$x^2 - 2$	119	142	167	194	223	254	287	322	359	398
prime factors	7,17	2,71	2,167	2,97	223	2,127	7,41	2,7,23	359	2,199

tells us that 2 is a square for the primes 2, 7, 17, 23, 31, 41, 47, 71, 79, 97, 127, 167, 199, 223, 359. Apart from 2 all the other moduli are $\equiv \pm 1$ (mod 8). So it looks as if 2 is a square *modulo p* when $p \equiv \pm 1$ (mod 8) and that 2 is a non-square *modulo p* when $p \equiv \pm 3$ (mod 8).

We shall leave a complete proof of the moduli for which 2 is a square to **50**. **Half-size products**. The historical motivation for **sections 48, 49, 50, 51** and **52** is described at the end of **section 53**.

49c Counting dots in a rectangle – comments

For $p = 13$ and $q = 7$ there are 6×3 lattice points in the rectangle. For $p = 13$ and $q = 17$ there are 48 lattice points in the rectangle. For $p = 11$ and $q = 7$ there are 15 lattice points in the rectangle.

The number of lattice points is

$$\left(\frac{p-1}{2}\right) \times \left(\frac{q-1}{2}\right)$$

The product is odd if and only if both factors are odd and since p and q are odd this happens when both p and $q \equiv 3$ *modulo* 4.

For each lattice point (x, y) the point $(\frac{1}{2}(p + 1) - x, \frac{1}{2}(q + 1) - y)$ is also a lattice point. Each of these pairs is joined by a line whose mid point is $(\frac{1}{4}(p + 1), \frac{1}{4}(q + 1))$.

The half turn

$$(x, y) \rightarrow \left(\frac{p+1}{2} - x, \frac{q+1}{2} - y\right)$$

maps the rectangle with vertices $QRST$ to itself since the corners are mapped $Q \leftrightarrow S$ and $R \leftrightarrow T$. More precisely

$$
\begin{array}{ccccccc}
& 1 & \leq & x & \leq & \tfrac{1}{2}(p-1) \\
\Leftrightarrow & -1 & \geq & -x & \geq & -\tfrac{1}{2}(p-1) \\
\Leftrightarrow & \tfrac{1}{2}(p+1)-1 & \geq & \tfrac{1}{2}(p+1)-x & \geq & \tfrac{1}{2}(p+1)-\tfrac{1}{2}(p-1) \\
\Leftrightarrow & \tfrac{1}{2}(p-1) & \geq & \tfrac{1}{2}(p+1)-x & \geq & 1
\end{array}
$$

Also since p and q are odd $\tfrac{1}{2}(p+1)$ and $\tfrac{1}{2}(q+1)$ are integers and so integer points will always be mapped to integer points.

If

$$
(x, y) = \left(\frac{p+1}{2} - x, \; \frac{q+1}{2} - y \right)
$$

then $(x, y) = (\tfrac{1}{4}(p+1), \tfrac{1}{4}(q+1))$, the centre of the half turn, which is only a lattice point when p and $q \equiv 3 \pmod 4$.

When $p = 11$, $q = 7$ we have an odd number, 15, of lattice points in the rectangle and the point $(3, 2) = (\tfrac{1}{4}(11+1), \tfrac{1}{4}(7+1))$ is an unpaired lattice point.

When $p = 7$, $q = 19$ we have an odd number, 27, of lattice points in the rectangle and the point $(2, 5) = (\tfrac{1}{4}(7+1), \tfrac{1}{4}(19+1))$ is an unpaired lattice point.

The matching of odd and even numbers of lattice points in a rectangle with the two cases in the law of quadratic reciprocity is due to Gauss' student Ferdinand Eisenstein in *Journal für die Reine und angewandte Mathematik* Vol. 27, 1844.

50c Half-size products – comments

	a	a	$2a$	$3a$	negative entries
	1	1	2	3	0
	2	2	−3	−1	2
mod 7	3	3	−1	2	1
	4	−3	1	−2	2
	5	−2	3	1	1
	6	−1	−2	−3	3

In each row the numbers are 1, 2, 3 in some order. The squares *modulo* 7 are 1, 2 and 4. Rows 1, 2 and 4 have an even number of negative entries. The non-squares *modulo* 7 are 3, 5 and 6 which lead the rows with an odd number of negative entries.

From this we conjecture that the number of negative entries is even/odd if the row number corresponds to a square/non-square *modulo* p. We can test the conjecture *modulo* 11 where the squares are 1, 3, 4, 5 and 9.

a	a	$2a$	$3a$	$4a$	$5a$	negative entries
1	1	2	3	4	5	0
2	2	4	−5	−3	−1	3
3	3	−5	−2	1	4	2
4	4	−3	1	5	−2	2
5	5	−1	4	−2	3	2
6	−5	1	−4	2	−3	3
7	−4	3	−1	−5	2	3
8	−3	5	2	−1	−4	3
9	−2	−4	5	3	1	2
10	−1	−2	−3	−4	−5	5

mod 11 (label at left of rows 5/6 area)

Here we have been considering some examples of sets

$$S_a[-\tfrac{p}{2}, \tfrac{p}{2}] \equiv \{a, 2a, 3a, \ldots, \tfrac{1}{2}(p-1)a\} \pmod{p} \text{ for hcf}(a, p) = 1$$

In each case that you have calculated, you will have noticed that each row of the table contains the numerical values $1, 2, \ldots, \tfrac{1}{2}(p-1)$ and they are all different. So it looks as if one number in $S_a[-\tfrac{p}{2}, \tfrac{p}{2}]$ is congruent to ± 1, another to $\pm 2, \ldots$, another to $\pm \tfrac{1}{2}(p-1) \pmod{p}$.

We can establish this by showing that a contradiction would result if two of these numbers were numerically indistinguishable *modulo p*. Suppose that $xa \equiv \pm ya \pmod{p}$; then

$$a(x \mp y) \equiv 0 \pmod{p}$$
$$x \mp y \equiv 0 \pmod{p}$$
$$x \equiv \pm y \pmod{p}$$

but since $1 \le x, y \le \tfrac{1}{2}(p-1)$ this means $x = y$, so that the product

$$a.2a. \ldots . \tfrac{1}{2}(p-1)a \equiv \pm(1.2.3. \ldots . \tfrac{1}{2}(p-1)) \pmod{p}$$

We can use this to prove the following result which was first proved by Gauss in 1798.

Gauss' lemma
Let p be a prime number and let hcf$(a, p) = 1$. If n denotes the number of negative integers in the set

$$S_a[-\tfrac{1}{2}p, \tfrac{1}{2}p] \equiv \{a, 2a, 3a, \ldots, \tfrac{1}{2}(p-1)a\} \pmod{p}$$

then a is a square *modulo p* if and only if n is even.

From the work above we know that

$$a.2a.3a. \ldots . \frac{1}{2}(p-1)a \equiv (-1)^n \left(\frac{p-1}{2} \right)! \pmod{p}$$

where there are n negative numbers in $S_a[-\tfrac{1}{2}p, \tfrac{1}{2}p]$.
    Hence

$$a^{\frac{1}{2}(p-1)} \equiv (-1)^n \pmod{p}$$

From **39. Powers of squares and non-squares** we know that a is a square *modulo* p if and only if $a^{\frac{1}{2}(p-1)} \equiv 1 \pmod{p}$.

The entries in the set

$$S_2 = \{2, 2.2, 3.2, \ldots, \tfrac{1}{2}(p-1).2\}$$

are all smaller than p so the only negative least residues will come from the numbers greater than $\frac{p}{2}$. That is, as soon as we are multiplying 2 by an integer greater than $\frac{p}{4}$. Hence

$$n = \frac{p-1}{2} - \left\lfloor \frac{p}{4} \right\rfloor$$

where $\lfloor x \rfloor$ denotes the integer such that $\lfloor x \rfloor \le x < \lfloor x \rfloor + 1$, so, for example, $\lfloor \pi \rfloor = 3$. $\lfloor x \rfloor$ is called the integer part of x or the integer floor of x.

If $p = 8k + 1$ then $n = \frac{1}{2}(8k) - \lfloor \frac{1}{4}(8k+1) \rfloor = 4k - 2k = 2k$

If $p = 8k + 3$ then $n = \frac{1}{2}(8k+2) - \lfloor \frac{1}{4}(8k+3) \rfloor = 4k + 1 - 2k = 2k + 1$

If $p = 8k + 5$ then $n = \frac{1}{2}(8k+4) - \lfloor \frac{1}{4}(8k+5) \rfloor = 4k + 2 - 2k - 1 = 2k + 1$

If $p = 8k + 7$ then $n = \frac{1}{2}(8k+6) - \lfloor \frac{1}{4}(8k+7) \rfloor = 4k + 3 - 2k - 1 = 2k + 2$

and using $2^{\frac{1}{2}(p-1)} \equiv (-1)^n \pmod{p}$ we discover that $2^{\frac{1}{2}(p-1)} \equiv 1 \pmod{p}$ for $p \equiv \pm 1 \pmod 8$ and $2^{\frac{1}{2}(p-1)} \equiv -1 \pmod{p}$ for $p \equiv \pm 3 \pmod 8$.

Hence 2 is a square *modulo* p if and only if $p \equiv \pm 1$ *modulo* 8.

51c From dashes to dots – comments

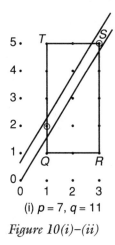 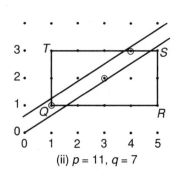

(i) $p = 7, q = 11$ (ii) $p = 11, q = 7$

Figure 10(i)–(ii)

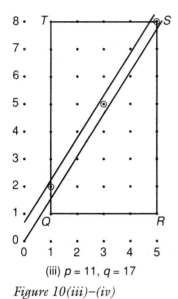
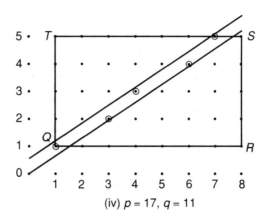

(iii) $p = 11$, $q = 17$ (iv) $p = 17$, $q = 11$

Figure 10(iii)–(iv)

	x	1	2	3
	qx	11	22	33
	qx/p	$^{11}/_7$	$^{22}/_7$	$^{33}/_7$
$p = 7$	$qx/p - \lfloor qx/p \rfloor$	$^4/_7$	$^1/_7$	$^5/_7$
$q = 11$	$< \frac{1}{2}, > \frac{1}{2}$	$>$	$<$	$>$
	numerically least residue $qx \bmod p$	-3	1	-2
	integer between qx/p and $qx/p + \frac{1}{2}$	*yes*	*no*	*yes*

	x	1	2	3	4	$5 = \frac{1}{2}(p-1)$
	qx	7	14	21	28	35
	qx/p	$^7/_{11}$	$^{14}/_{11}$	$^{21}/_{11}$	$^{28}/_{11}$	$^{35}/_{11}$
$p = 11$	$qx/p - \lfloor qx/p \rfloor$	$^7/_{11}$	$^3/_{11}$	$^{10}/_{11}$	$^6/_{11}$	$^2/_{11}$
$q = 7$	$< \frac{1}{2}, > \frac{1}{2}$	$>$	$<$	$>$	$>$	$<$
	numerically least residue $qx \bmod p$	-4	3	-1	-5	2
	integer between qx/p and $qx/p + \frac{1}{2}$	*yes*	*no*	*yes*	*yes*	*no*

	x	1	2	3	4	$5 = \frac{1}{2}(p-1)$
	qx	17	34	51	68	85
	$\frac{qx}{p}$	$\frac{17}{11}$	$\frac{34}{11}$	$\frac{51}{11}$	$\frac{68}{11}$	$\frac{85}{11}$
$p = 11$	$\frac{qx}{p} - \lfloor\frac{qx}{p}\rfloor$	$\frac{6}{11}$	$\frac{1}{11}$	$\frac{7}{11}$	$\frac{2}{11}$	$\frac{8}{11}$
$q = 17$	$<\frac{1}{2}, >\frac{1}{2}$	>	<	>	<	>
	numerically least residue $qx \bmod p$	−5	1	−4	2	−3
	integer between $\frac{qx}{p}$ and $\frac{qx}{p} + \frac{1}{2}$	yes	no	yes	no	yes

	x	1	2	3	4	5	6	7	$8 = \frac{1}{2}(p-1)$
	qx	11	22	33	44	55	66	77	88
	$\frac{qx}{p}$	$\frac{11}{17}$	$\frac{22}{17}$	$\frac{33}{17}$	$\frac{44}{17}$	$\frac{55}{17}$	$\frac{66}{17}$	$\frac{77}{17}$	$\frac{88}{17}$
$p = 17$	$\frac{qx}{p} - \lfloor\frac{qx}{p}\rfloor$	$\frac{11}{17}$	$\frac{5}{17}$	$\frac{16}{17}$	$\frac{10}{17}$	$\frac{4}{17}$	$\frac{15}{17}$	$\frac{9}{17}$	$\frac{3}{17}$
$q = 11$	$<\frac{1}{2}, >\frac{1}{2}$	>	<	>	>	<	>	>	<
	numerically least residue $qx \bmod p$	−6	5	−1	−7	4	−2	−8	3
	integer between $\frac{qx}{p}$ and $\frac{qx}{p} + \frac{1}{2}$	yes	no	yes	yes	no	yes	yes	no

In each of the four tables, there is an integer between $\frac{qx}{p}$ and $\frac{qx}{p} + \frac{1}{2}$ precisely when the numerically least residue $\equiv qx \pmod{p}$ is negative.

Since the vertical gap between the two lines is $\frac{1}{2}$ and the lattice points are 1 apart there can be at most one lattice point between the lines for each value of x.

From the figure below it is clear that if the integer part of $\frac{qx}{p}$ is more than $\frac{1}{2}$ below the line $y = \frac{qx}{p}$, then the next lattice point above the line will be below the line $y = \frac{qx}{p} + \frac{1}{2}$.

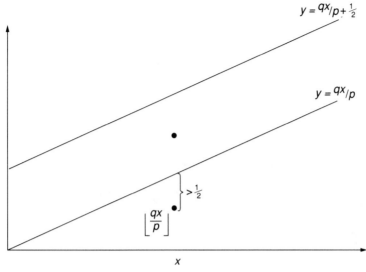

Figure 11

To show that the lattice points in the region are linked with the negative entries in S,

$$\frac{qx}{p} - \left\lfloor \frac{qx}{p} \right\rfloor > \frac{1}{2} \Leftrightarrow \left\lfloor \frac{qx}{p} \right\rfloor + 1 > \frac{qx}{p} > \left\lfloor \frac{qx}{p} \right\rfloor + \frac{1}{2}$$

$$\Leftrightarrow p \left\lfloor \frac{qx}{p} \right\rfloor + p > qx > p \left\lfloor \frac{qx}{p} \right\rfloor + \frac{p}{2}$$

$$\Leftrightarrow qx \equiv t \bmod p \text{ where } p > t > \frac{p}{2}$$

Since q and p are coprime there are no values of x for $1 \le x \le \frac{1}{2}(p-1)$ which give lattice points on the lines $y = \frac{q}{p}x$ or $y = \frac{q}{p}x + \frac{1}{2}$.

The number of lattice points in the region A equals the number of negative numerically least residues $\equiv qx \pmod p$, and this number is even when q is a square *modulo p*, and odd when q is a non-square *modulo p* (from **50. Half-size products**).

The argument here is due to Gauss (1808) and the geometry to which it corresponds to Eisenstein (1844).

52c Quadratic reciprocity – comments

The mirror line in which the (p, q) rectangle is reflected onto the (q, p) rectangle is $y = x$. Under this reflection $(x, y) \rightarrow (y, x)$, so lattice points always map onto lattice points. The reflection of $y = qx/p$ in $y = x$ is $x = qy/p$, or $y = px/q$. The reflection of $y = qx/p + \frac{1}{2}$ in $y = x$ is $x = qy/p + \frac{1}{2}$ or $y = px/q - p/2q$.

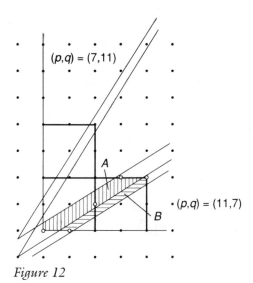

Figure 12

7 is a non-square *modulo* 11 so $(7 \mid 11) = -1$ and from **section 51**, the number of lattice points in A is odd. 11 is a square *modulo* 7 so $(11 \mid 7) = 1$ and from **section 51**, the number of lattice points in B is even.

Under the half turn $(x, y) \to (\frac{1}{2}(p + 1) - x, \frac{1}{2}(q + 1) - y)$, the point $(0, \frac{1}{2}) \to (\frac{1}{2}(p + 1), \frac{1}{2}q)$ and the point $(\frac{1}{2}p, \frac{1}{2}(q + 1)) \to (\frac{1}{2}, 0)$ so the line $y = qx/p + \frac{1}{2}$ is mapped to the line $x = py/q + \frac{1}{2}$. The boundaries of $A \cup B$ are mapped to themselves by the half turn, and so the region is mapped to itself.

If the number of lattice points in $A \cup B$ is odd, then the number in A is odd and the number in B is even, or vice versa. But an odd number in A implies $(q \mid p) = -1$ and an even number in B implies $(p \mid q) = 1$, from **section 51**, so $(p \mid q) \neq (q \mid p)$.

If the number of lattice points in $A \cup B$ is even, then the numbers of lattice points in A and B are both even or both odd, giving either $(p \mid q) = (q \mid p) = 1$ or $(p \mid q) = (q \mid p) = -1$.

But we have found that $A \cup B$ has an odd number of lattice points when $QRST$ has an odd number of lattice points, which is when $\frac{1}{2}(p - 1)\frac{1}{2}(q - 1)$ is odd. This only happens when $p \equiv q \equiv 3 \pmod 4$. So $(p \mid q) \neq (q \mid p)$ when $p \equiv q \equiv 3 \pmod 4$.

Otherwise $(p \mid q) = (q \mid p)$. The conjecture of **48. Primes as squares and non-squares** has been established.

This profound and remarkable result (the **law of quadratic reciprocity**) is due to Gauss (1799) and enables us to calculate directly whether a given number is or is not a square *modulo* a particular prime. We illustrate by proving that 1021 is a non-square *modulo* 257.

$$
\begin{aligned}
(1021 \mid 257) &= (250 \mid 257) && \text{because } 1021 \equiv 250 \ (\text{mod } 257) \\
&= (25 \mid 257)(10 \mid 257) && \text{because } (ab \mid p) = (a \mid p)(b \mid p) \\
&= (10 \mid 257) && \text{because } 25 \text{ is a square} \\
&= (2 \mid 257)(5 \mid 257) && \\
&= (5 \mid 257) && \text{because } 257 \equiv 1 \ (\text{mod } 8) \ (\text{see \textbf{section 50}}) \\
&= (257 \mid 5) && \text{by quadratic reciprocity} \\
&= (2 \mid 5) && \text{because } 257 \equiv 2 \ (\text{mod } 5) \\
&= -1 && \text{because } 5 \equiv -3 \ (\text{mod } 8) \ (\text{see \textbf{50}})
\end{aligned}
$$

53c Adding squares – comments

For x and $y \leq 10$, the primes of the form $x^2 + y^2$ in the table of **section 35** are 2, 5, 13, 17, 29, 37, 41, 53, 61, 73, 89, 97, 101, 109, 113, 149, 181. Apart from 2, each of these primes is $\equiv 1 \pmod 4$. If $x^2 + y^2 = p$, then $x^2 + y^2 \equiv 0 \pmod p$ and so $x^2 \equiv -y^2 \pmod p$. From **38. Squares and non-squares** this is only possible if -1 is a square, which needs $p \equiv 1 \pmod 4$ from **33. Square roots of -1**. In fact the primes in this table are precisely 2 and the primes $\equiv 1 \pmod 4$. The fact that they all appear if only the table is extended far enough, is a rather subtle deduction from the existence of an equation of the form $x^2 + y^2 = 0 \pmod p$, when $y = 1$, for the right kinds of primes p. The equation $(x^2 + y^2)(a^2 + b^2) = (xa - yb)^2 + (xb + ya)^2$ (used in **section 36**) shows that the product of two sums of squares is again a sum of squares, but it can also be used to show how an equation of the kind $x^2 + y^2 = mp$ leads to an equation of the kind $x^2 + y^2 = p$. In fact the numbers in the table may have any prime factors whatever, but primes $\equiv 3 \pmod 4$ always appear to an even power.

The primes in the first table are 2, 3, 11, 17, 19, 41, 43, 59, 67, 73, 83, 89, 97, 107, 113, 131, 137, 163, 179, 211, 281. Apart from 2, these are $\equiv 1$ or $3 \pmod 8$. If $x^2 + 2y^2 = p$, then $x^2 + 2y^2 \equiv 0 \pmod p$ and so $x^2 \equiv -2y^2 \pmod p$. Assume $p > 2$. From **38. Squares and non-squares** this congruence is only possible if -2 is a square, which needs either both -1 and 2 to be squares or neither -1 nor 2 to be squares. The first condition gives $p \equiv 1 \pmod 8$ and the second gives $p \equiv 3 \pmod 8$ using **section 33** and **section 49**. Futher analysis depends on using the equation $(x^2 + 2y^2)(a^2 + 2b^2) = (xa - 2yb)^2 + 2(xb + ya)^2$, which incidentally shows that any product of reachable primes is reachable. Primes $\equiv 5$ or $7 \pmod 8$ only appear to an even power as factors of numbers in the table.

The primes in the second table are 3, 7, 13, 19, 31, 37, 43, 61, 67, 73, 79, 97, 103, 109, 127, 139, 151, 157, 163, 193, 211, 241, 307, 349. Apart from 3, these are $\equiv 1 \pmod 6$. If $x^2 + 3y^2 = p$, then $x^2 + 3y^2 \equiv 0 \pmod p$ and so $x^2 \equiv -3y^2 \pmod p$. Assume $p > 3$. From **38. Squares and non-squares** this congruence is only possible if -3 is a square, which needs either both -1 and 3 to be squares or neither -1 nor 3 to be squares. Now -1 is a square when $p \equiv 1 \pmod 4$ (from **section 33**) and 3 is a square when $(3 \mid p) = 1$. But when $p \equiv 1 \pmod 4$, $(3 \mid p) = (p \mid 3)$ (using **52. Quadratic reciprocity**) and $(p \mid 3) = 1$ when $p \equiv 1 \pmod 3$. These two conditions give $p \equiv 1 \pmod{12}$. Moreover, -1 is a non-square when $p \equiv 3 \pmod 4$, and 3 is a non-square when $(3 \mid p) = -1$. When $p \equiv 3 \pmod 4$, $(3 \mid p) = -(p \mid 3)$ (using **52. Quadratic reciprocity**) and as before $-(p \mid 3) = -1$ when $p \equiv 1 \pmod 3$. These two conditions give $p \equiv 7 \pmod{12}$. $p \equiv 1$ or $7 \pmod{12}$ if and only if $p \equiv 1 \pmod 6$.

Futher analysis depends on using the equation $(x^2 + 3y^2)(a^2 + 3b^2) = (xa - 3yb)^2 + 3(xb + ya)^2$, which incidentally shows that any product of the reachable primes is reachable. Primes $\equiv 5$ or $11 \pmod{12}$ or equivalently $\equiv 5 \pmod 6$ only appear to an even power as factors of numbers in the table.

These proofs are incomplete. They show that primes $\equiv 3 \pmod 4$ may not be expressed in the form $x^2 + y^2$, but they do not establish that every prime $\equiv 1 \pmod 4$ may be so expressed. They show that primes $\equiv 5$ or $7 \pmod 8$ may not be expressed in the form $x^2 + 2y^2$, but they do not establish that every prime $\equiv 1$ or $3 \pmod 8$ may be so expressed. They show that primes $\equiv 5 \pmod 6$ may not be expressed in the form $x^2 + 3y^2$, but they do not establish that every prime $\equiv 1 \pmod 6$ may be so expressed.

To show how these proofs may be completed, we will illustrate the argument in this last case before giving the general proof for this case. It is based on knowing that all primes $\equiv 1 \pmod 6$ up to a certain value are expressible in the required form, and using this knowledge to express the next one up.

The prime number 337 did not appear in our second table but is $\equiv 1 \pmod 6$. So there is a solution of $x^2 \equiv -3 \pmod{337}$, and in fact a solution with $|x| < 337/2$. This is $80^2 \equiv -3 \pmod{337}$ and this gives $80^2 + 3 = 6403 = 19 \cdot 337$. It is of critical importance here that $19 < 337$ and that the prime factors of 19 are $\equiv 1 \pmod 6$ and so have a representation of the desired kind. In fact $19 = 4^2 + 3 \cdot 1^2$.

So

$$19 \cdot 337 \cdot 19 = (80^2 + 3 \cdot 1^2)(4^2 + 3 \cdot 1^2) = (80 \cdot 4 + 3 \cdot 1 \cdot 1)^2 + 3(80 \cdot 1 - 4 \cdot 1)^2$$

and

$$19^2 \cdot 337 = 323^2 + 76^2 = 19^2 \cdot 17^2 + 19^2 \cdot 3 \cdot 4^2$$

so

$$337 = 17^2 + 3 \cdot 4^2$$

For the general proof, suppose that all the primes $\equiv 1 \pmod 6$ less than p have been shown to be representable in the form $x^2 + 3y^2$ and that $p \equiv 1 \pmod 6$. Then there exists a solution of $x^2 \equiv -3 \pmod p$, with $|x| < p/2$. This shows that $x^2 + 3 = m \cdot p$ for some $m < p$. Now $x^2 \equiv -3 \pmod m$, so none of the prime factors of $m \equiv 5 \pmod 6$. Let q be a prime factor of m; then $q \equiv 1 \pmod 6$ and $q < p$, so q is representable in the form $x^2 + 3y^2$. Let $q = a^2 + 3b^2$. Then, with $y = 1$,

$$mpq = (x^2 + 3y^2)(a^2 + 3b^2) = (xa - 3yb)^2 + 3(xb + ya)^2 = (xa + 3yb)^2 + 3(xb - 3ya)^2$$

Now $a^2 \equiv -3b^2 \pmod q$ and $x^2 \equiv -3y^2 \pmod q$, so $x^2a^2 \equiv 9y^2b^2 \pmod q$ and $xa \equiv \pm 3yb \pmod q$.

If $xa \equiv 3yb$ then $x^2a \equiv 3xyb$, so $-3y^2a \equiv 3xyb$, and $-ay \equiv xb \pmod q$. If $xa \equiv -3yb$ then $x^2a \equiv -3xyb$, so $-3y^2a \equiv -3xyb$, and $ay \equiv xb \pmod q$. So either $xa - 3yb$ and $xb + ya$ have a factor q, or $xa + 3yb$ and $xb - ya$ have a factor q. In the first case we use $mpq = (xa - 3yb)^2 + 3(xb + ya)^2$ to obtain

$$\left(\frac{m}{q}\right)p = \left(\frac{xa - 3yb}{q}\right)^2 + 3\left(\frac{xb + ya}{q}\right)^2$$

which is an equation in integers, expressing a lesser multiple of p in the required form. In the second case the alternative equality contains terms divisible by q. The argument with q may now be applied to the other prime factors of m until the representation of p has been obtained.

Historically, the study of quadratic reciprocity, conjectured in **section 48** and proved in the sequence **49, 50, 51** and **52**, was motivated by the desire to determine the integers expressible in the form $x^2 + ny^2$, which because of the equation $(x^2 + ny^2)(a^2 + nb^2) = (xa - nyb)^2 + n(xb + ya)^2$, was seen to reduce to the determination of primes which may be so expressed. As we have seen this depends upon whether or not $-n$ is a square for the particular prime as a modulus, and the determination of this, as has emerged in our study of the third table, needs quadratic reciprocity.

While Fermat claimed the results we have obtained for n = 1, 2 *and* 3 *in 1654, the first proofs are due to Euler in 1742 and 1744. Legendre tried to generalise these results (1785 and 1798) and believed he had a proof of quadratic reciprocity. The first accepted proof of quadratic reciprocity is due to Gauss (1799).*

Index

A reference to 16 is to Section 16 in the first half of the book. A reference to 16c is to the Comments on Section 16 to be found in the second half of the book.

al-Khowarizmi   1c

Binary numbers (bicimals)   2, 2c

Caesar code   24, 24c
cancellation law   30c, 31c, 43c
Cardan   4c
Chinese remainder theorem   21, 21c, 34c
code
   Caesar   24, 24c
   public key system   47
common factors   5
composite numbers   4, 4c, 13c, 14, 14c
congruence   5, 5c, 6, 13, 16, 17c, 19, 21, 23, 23c
   simultaneous   21, 21c
contradiction, proof by   13c, 15c
coprime numbers   22c, 23c
Crelle   39c

Decimal
   recurring   45, 45c, 46, 46c
   terminating   45, 45c
Diophantus   37c
divisibility
   by 2, 3, 4, 5, 6, 7, 8, 10, 11   11, 11c
   by 9   1, 1c, 11, 11c
division algorithm   5c, 6c, 17c, 18c
divisors
   coprime   9c
   lattice of   4
   number of   4, 4c
   prime see prime factors
Eisenstein   49c, 51c
Eratosthenes   13, 13c
Euclid   6c, 15c, 37c
Euclidean algorithm   6, 6c, 43c
Euler   22c, 31c, 39c, 44c, 53c
Euler's φ function   22, 22c, 24c, 40, 40c, 42, 42c, 44c, 46, 46c, 47, 47c
   multiplicative property   22c

factorial   32, 32c
factors   3c, 4, 4c, 5, 5c, 6, 6c, 7, 7c, 8, 8c, 13c, 14, 14c, 15, 15c, 18, 18c, 40, 43, 45, 45c.
Fermat   31c, 53c
Fermat's factorisation   28, 28c

Fermat's theorem
   mod 5, 7, 13   29, 29c
   mod 7   30, 30c with proof
   mod 3, 5, 11   31, 31c with proof
   mod p   31c, 32c, 39c, 42, 42c, 46, 46c
Fermat–Euler theorem   44c, 46, 46c, 47, 47c
Fibonacci   7, 7c, 18, 18c, 20, 20c
Fibonacci sequence   7, 7c
fundamental theorem of arithmetic   3, 3c, 4c,
   proof   9c

Gauss   22c, 23c, 38c, 40c, 42c, 50c, 51c, 52c, 53c
Gauss' law of quadratic reciprocity   48c
Gauss' lemma   50c
generators
   for addition   22c, 23c, 41, 41c
   for multiplication   41, 41c, 42, 42c

highest common factors   5, 5c, 6, 6c, 7, 7c, 9, 10c, 18, 40, 40c
hcf(a, b) as ax + by   6, 6c, 9, 9c, 10

identity   23, 32, 32c
induction   20, 20c
infinity of primes   13c, 15, 15c
inverse   23, 32, 32c

Keller   14c
Kersey   4c

Lagrange   32c, 39c
Lambert   45c
lattice of factors   4
Legendre   53c
   function   52, 52c
Leibniz   45c

modular arithmetic
   coprime moduli   21, 21c
   generators for addition   22, 22c, 23c
   generators for multiplication   41, 41c, 42, 42c
   mod 4   16, 16c, 17, 17c
   mod 10   12
   mod 12   17, 17c

modular arithmetic cont'd
  mod n 17c, 23c
  multiplication 23, 23c, 43, 43c, 44, 44c
multiplicative property 22c
multiplication 1, 1c, 12

non-squares 25, 25c, 38, 38c, 42, 42c, 48, 48c
non-transitive dice 19

order of an integer (mod n) 42, 42c

palindrome 1, 1c
Pascal 20c
Pisano 11c
place value 1, 1c, 2, 2c, 11, 11c, 12
polynomial congruence 39c
powers 29, 29c, 30, 30c, 31, 31c
prime factors 3, 3c, 4, 4c, 8, 8c, 13, 13c
  of numbers up to 100 4c
  of products 8, 8c
    proof 9c
  quantity of 13c
prime numbers 3, 3c, 4, 4c, 8, 8c, 9, 9c, 13, 13c
  as sums of squares 35c, 53, 53c
  infinity of 15, 15c
  pairs of 14, 14c
primitive root 42c
principle of mathematical induction 20, 20c
proof
  by contradiction 13c, 15c
  by exhaustion 12
  by induction 9c, 20, 20c
public key system 47, 47c
Pythagorean triples 37, 37c
  primitive 37, 37c

quadratic reciprocity 48, 48c, 52, 52c, 53c
quadratic residues *see* square numbers

recurring decimals 45, 45c, 46, 46c
relations: reflexive, symmetric, transtive 17c, 19, 19c
remainders 5c, 17c, 21
residue class 17c
Rhind papyrus 2c
RSA system 47, 47c

shift transformation 24
square-free 4c
squares and non-squares 25, 25c, 26, 26c, 27, 27c, 38, 38c, 39, 39c, 48, 48c, 50, 50c
  closure of 38, 38c
square numbers 25, 25c, 26, 26c, 27c, 28c, 33, 33c, 35, 36, 48, 48c, 49, 49c
  mod $3, 4, 5$ 16, 26, 26c, 37, 37c
  mod $5, 7$ 27, 27c
  mod 8 37, 37c
  -1 33, 33c, 34, 34c, 53c
  2 53c
square roots
  of 1 32, 32c
  of -1 33, 33c, 34, 34c
sums of squares
  in two ways 36, 36c
  $x^2 + y^2$ 35, 35c, 53, 53c
  $x^2 + 2y^2$ 53, 53c
  $x^2 + 3y^2$ 53, 53c
  $x^2 + ny^2$ 53, 53c

totient function 22, 22c

Wallis 45c, 46c
Waring 32c
Wilson 32c

$\mathbf{Z}$ 17c
$\mathbf{Z}_n$ 12, 17c, 22c
zero products 43, 43c